The Biology of
HEALTH&DISEASE

prepared by the Open University's
U205 Course Team

OPEN UNIVERSITY PRESS
Milton Keynes · Philadelphia

The U205 Course Team

U205 is a course whose writing and production has been the joint effort of many hands, a 'core course team', and colleagues who have written on specific aspects of the course but have not been involved throughout; together with editors, designers, and the BBC team.

Core Course Team

The following people have written or commented extensively on the whole course, been involved in all phases of its production and accept collective responsibility for its overall academic and teaching content.

Steven Rose (neurobiologist; course team chair; academic editor; Book VI coordinator)

Nick Black (community physician; Book IV coordinator)

Basiro Davey (immunologist; course manager; Book V coordinator)

Alastair Gray (health economist; Book III coordinator)

Kevin McConway (statistician; Book I coordinator)

Jennie Popay (social policy analyst; Book VIII coordinator)

Jacqueline Stewart (managing editor)

Phil Strong (medical sociologist; academic editor; Book II coordinator)

Other authors

The following authors have contributed to the overall development of the course and have taken responsibility for writing specific sections of it.

Lynda Birke (ethologist; author, Book V)

Eric Bowers (parasitologist; staff tutor)

David Boswell (sociologist; author, Book II; Book VII coordinator)

Eva Chapman (psychotherapist; author, Book V)

Andrew Learmonth (geographer; course team chair 1983; author, Book III)

Rosemary Lennard (medical practitioner; author, Books IV and V)

Jim Moore (historian of science; author, Book II)

Sean Murphy (neurobiologist; author, Book VI)

Rob Ransom (developmental biologist; author, Book IV)

George Watts (historian; author, Book II)

The following people have assisted with particular aspects or parts of the course.

Sylvia Bentley (course secretary)

Steve Best (illustrator)

Sheila Constantinou (BBC production assistant)

Ann Hall (indexer)

Rachel Hardman (designer)

Mark Kesby (illustrator)

Liz Lane (editor)

Vic Lockwood (BBC producer)

Laurie Melton (librarian)

Sue Walker (editor)

Peter Wright (editor)

External consultants

Alex Clarke (psychologist) Royal Free Hospital School of Medicine.

Tim Halliday (evolutionary biologist) Senior Lecturer in Biology, The Open University.

Marie Johnston (clinical psychologist) Royal Free Hospital School of Medicine.

Marshall Stark (medical microbiologist) Welsh National School of Medicine.

Harford Williams (parasitologist) Director of the Open University in Wales.

External Assessors

Course Assessors

Alwyn Smith President, Faculty of Community Medicine of the Royal Colleges of Physicians; Professor of Epidemiology and Social Oncology, University of Manchester.

Book IV Assessors

Lewis Wolpert Professor of Biology as Applied to Medicine, Middlesex Hospital Medical School.

Peter Abrahams Senior Lecturer in Anatomy, Middlesex Hospital Medical School.

Acknowledgements

The course team wishes to thank the following for their advice and contributions:

Sheila Adam (community physician) North West Thames Regional Health Authority.

John Ashton (community physician) Department of Community Health, University of Liverpool.

Open University Press, Celtic Court, 22 Ballmoor, Buckingham, MK18 1XW.

First published 1985. Reprinted 1990. Copyright © 1985 The Open University.

Designed by the Graphic Design Group of the Open University.

Typeset by the Pindar Group of Companies, Scarborough, North Yorkshire.
Printed in Great Britain by St Edmundsbury Press Limited, Bury St Edmunds, Suffolk

ISBN 0 335 15053 5

This book forms part of the Open University Press 'Health and Disease' series.
The complete list of books in the series is printed on the back cover.

Further information on Open University courses may be obtained from the Admissions Office, The Open University, P.O. Box 48, Walton Hall, Milton Keynes, MK7 6AB.

About this book

A note for the general reader

The Biology of Health and Disease is the fourth of a series of books on the subject of health and disease. It is designed so that it can be read on its own, like any other textbook, or studied as part of U205 *Health and Disease*, a second-level course for Open University students. As well as the eight textbooks and a Course Reader*, the course consists of eleven television programmes and five audiocassettes, plus various supplementary materials.

For Open University students, there is an *Introduction and Guide* to the course, which sets out a study plan for the year's work. This is supplemented where appropriate in the text by more detailed 'study comments', which you will find boxed, for easy reference, at the beginning of chapters. Also in the text you will find references to articles in the Course Reader. It is quite possible to follow the argument without reading these articles, although your understanding will be enriched if you do. Major learning objectives are listed at the end of each chapter along with questions that allow students to assess how well they are achieving these objectives. The index includes key words (in bold type) which can be looked up easily as an aid to revision. There is also a 'further reading' list for those who wish to pursue certain aspects of study beyond the limits of this book.

A guide for OU students

The Biology of Health and Disease examines the biological aspects of diseases — their immediate cause and the changes that occur in the structure and function of the human body that give rise to the symptoms and signs of disease. As we do not assume that the reader has any detailed knowledge of human biology, the first half of the book is devoted to providing an account of the basic biology of cells, tissues and organs, and the normal functioning of the body. It draws on the biological methods of investigation discussed in other books in the course and provides a foundation for many of the discussions of specific diseases in the second half of the book.

The book contains seventeen chapters, which fall into three main parts. Chapters 2 to 4 are concerned with the biochemistry, cell biology and development of the human body. This covers the basic information on how cells and tissues function and enables the reader to understand the explanations of the normal functioning (physiology) of organs and body systems in Chapters 5 to 10. The rest of the book (Chapters 11 to 17) is concerned with diseases. After a discussion of why disease occurs, in Chapter 11, a wide range of diseases, categorised by cause, is considered. The book ends with a brief discussion of the contribution of biomedicine to the occurrence of disease.

The time allowed for studying Book IV is four and a half weeks or 45 hours, with an extra half-week for revision. The table overleaf gives a more detailed breakdown to help you to pace your study. You need not follow it slavishly, but do not allow yourself to fall behind. If you find a section of the work difficult, do what you can at this stage and then rework the material at the end of your study of this book.

---

* Black, Nick *et al.* (eds) (1984) *Health and Disease: A Reader*, The Open University Press.

Study time for Book IV (total 45 hours)

Chapter	Time/hours	Course Reader	Time/hours	TV and audiocassettes	Time/hours
1	1				
2	$3\frac{1}{4}$ } $8\frac{1}{2}$				
3	$2\frac{1}{4}$				
4	2			TV 4 *Life Before Birth*	$\frac{3}{4}$
5	1				
6	3 } $8\frac{1}{4}$				
7	$1\frac{1}{2}$				
8	$2\frac{3}{4}$				
9	2				
10	$1\frac{1}{2}$ } $8\frac{1}{2}$				
11	3				
12	2				
13	4				
14	$4\frac{1}{2}$ } 10	Strassburg (1982)	$\frac{3}{4}$		
15	$1\frac{1}{2}$				
16	$3\frac{1}{2}$ } $4\frac{1}{2}$				
17	1	Beeson (1980)	$\frac{3}{4}$		

Assessment TMA 04 (3 hours) should be completed at the end of your study of Book IV.

As a rough guide to pacing your work, we suggest that you study Chapters 1 to 4 in the first week, 5 to 8 in the second, and so on.

Contents

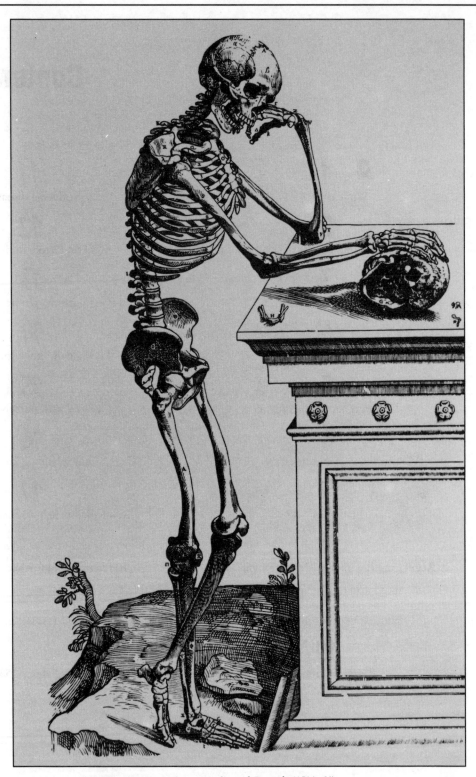

Lateral view of the skeleton by Andreas Vesalius of Brussels (1514–64).

1
What
is
life?

Health depends on the body maintaining its internal harmony. When that harmony is so disturbed that the body is unable to respond adequately to compensate for the disruption, then disease occurs. Given the vast number of potentially disruptive factors that humans face every day, from conception, through infancy and childhood, to adulthood and old age, it seems miraculous that for the most part people remain free of disease, a testimony to the body's extraordinary powers of compensation. The subject of this book is the normal functioning that maintains the human body in a healthy state from birth to maturity and the abnormalities that result when that harmony is disrupted.

The ability of the body to maintain the harmony of its internal environment is known to biologists as *homeostasis*. Its ability to grow and change over time is known as *development*. Both processes operate at many different levels, from the continuous chemical reactions that take place in all the cells of the body to the mechanisms that control the long-term processes of growth and ageing. Not only the internal environment has to be maintained in a state that is steady, while simultaneously capable of change and development, but of equal importance is the maintenance of harmony with the external environment —with other species, such as bacteria and viruses, with physical factors, such as temperature, and with other people. Health involves such psychological and emotional matters as well as simple relationships with the physical world. This need to maintain both internal and external harmony is, of course, not unique to humans, but is a feature of all living organisms. Our starting point therefore must be a consideration of what is meant by 'living', in contrast to 'non-living'. What are the common, essential properties shared by all living organisms?

There are six main properties which set living organisms apart from non-living matter. Organisms *move* independently of external forces; they *feed*, that is, they absorb material from the external environment and return waste products to the environment; they *respond* actively to changes in their environment and in doing so change that environment; they *grow* to adult; they *repair* themselves when damaged, but eventually undergo a puzzling transformation from life to the inanimate state of death; and they *reproduce* themselves either by budding off apparently identical copies or by sexual reproduction.

All living creatures have definite *forms*—by which is meant simply their physical structure. The surface membrane of the microorganism, the petal of an orchid, the stripes of a zebra, and the characteristic five fingers of the human hand are all examples of biological form. But, unlike the form of a crystal or of a cloud, biological form is regarded as intimately associated with *function*, that is, with how the structure works. Thus a biologist looks at the surface membrane of a microorganism and interprets it as the interface between organism and environment, keeping the two apart yet capable of changing shape so as to move the organism within the environment. The flower of the orchid is attractive to insects because of its colour or scent, and insects landing on it pick up pollen, which is then transferred from flower to flower. In this way the orchid's petal functions within the plant's reproductive process.

    □   What might the function of the zebra's stripes be?
    ■   They act as a sort of camouflage, making the zebra harder to see by predators, such as lions.
    □   And what might be the function of the human hand?
    ■   The hand is *adapted* to hold, grasp and manipulate objects. The word 'adapted' is a key biological term, a measure of the fitness of form for function. Because of the particular relation of the thumb to the other four fingers (each finger can be touched by the thumb), there is an extraordinary flexibility between precision, gentleness and strength in the hand's grasp.

It may be apparent to you that, given the number of different levels at which biological events take place (from chemicals and cells, through to whole organisms and groups of organisms), there are also several perspectives that biologists can adopt to investigate the processes involved. For example, study of the relationship of form to function is the domain of *physiology*. Let us consider for a moment a typical question that might be asked by a biologist, and see the number of different ways it might be answered, depending on the biological discipline of the respondent. Consider the question of why zebras are striped.

To answer this a cell biologist might choose to look at the zebra's hide under the microscope. This would reveal that the stripes are composed of hairs of different colours. Each hair grows from a particular unit in the zebra's skin, a *cell*. Indeed, one of the principles of the organisation of living forms is that all are composed of cells, small box-like structures a few thousands or hundredths of a millimetre

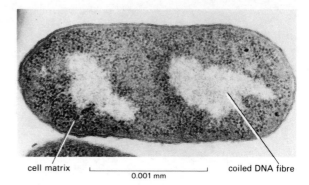

cell matrix |————————| coiled DNA fibre
                 0.001 mm

Figure 1.1 Electron micrograph of a bacterium.

across. This principle of the cellular organisation of life was recognised and formulated by the early microscopists in the middle of the nineteenth century.

Higher magnification reveals that cells themselves have internal structures. Look for example at either Plate I (a liver cell) or Figure 1.1, which shows a bacterial cell. Most of the interior of the cell is composed of water, but dissolved in the water are many different substances. The cells in the zebra skin would in addition contain chemicals capable of manufacturing hair, and some of these hair-making cells would also make pigment to colour the hair black. These chemicals are studied by biochemists.

So if we ask why the zebra is striped, one answer would be: because some of the cells in its skin contain the machinery for making black hairs and these cells are arranged in a striped pattern. That is, we have explained the *form* of the zebra's stripes in terms of the properties of the cells and molecules which compose the zebra's skin.

    □   What might you call this type of explanation?
    ■   It is a *reductionist* explanation: the properties of the organism are described in terms of its constituent parts. The biological disciplines which try to provide such explanations are those of *cell biology* and *biochemistry*, the study of the chemical nature of living processes.

Note that this description of the zebra's stripes has also revealed one of the other characteristic properties of multicellular living organisms, that of *organisational hierarchy* (Figure 1.2). Chemicals are grouped into subcellular structures (for example, the nucleus) which are themselves organised into cells; the cells are grouped into *tissues* (such as the zebra skin) and *organs* (such as the brain, liver or stomach); the organs and tissues themselves compose the *organism*. Biologists study the organism at all these different hierarchical levels.

In addition, biologists study societies of organisms in their environment and in their interaction with other organisms. Ethology is concerned with the social behaviour of animals and *ecology* with the relationships between

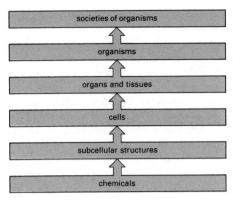

Figure 1.2 The hierarchy of biological organisation.

groups of living creatures co-existing in any particular environment. All living creatures, even the simplest, interact dynamically with their environment, constantly changing it, and themselves, in the process. Whereas the reductionist disciplines, like biochemistry, give a sort of 'bottom up' explanation of the zebra's stripes, the sciences like ecology and ethology are more *holistic* or 'top down' and tend to explore the relationship between the whole animal — including its stripes — and its environmental context.

Even this does not exhaust the questions that a biologist may ask about the zebra's stripes. It is not enough to know that they exist now and are composed of such-and-such a set of cells containing such-and-such a set of chemicals. A zebra begins life, like any other animal, by the fusion of an ovum and a sperm cell. A rapidly dividing ball of cells begins to *differentiate* into an embryo containing the characteristic features of the zebra. It turns out that the pattern of its stripes is laid down during this development, by the sequence in which particular cells are instructed, at particular times, to make or not make the dark pigment. How this comes about is the province of the disciplines of

genetics and *developmental biology*, which recognise that the present features of the zebra cannot be understood without knowing how the animal has grown and developed.

Zebras have parents, which were striped in their turn. And they had parents too. The present-day zebra is the result of the *evolution* of biological forms over thousands of generations and millions of years. The key to the evolutionary process, as offered by the naturalists Charles Darwin and Alfred Russel Wallace in 1859, is a very simple one. All living creatures tend to resemble their parents, with minor variations. There are not enough resources in the environment to support all offspring, and so not all creatures which are born live long enough to reproduce. The ones which are more likely to live are those that are best adapted to the environment.

Suppose the zebra's stripes are a form of camouflage that helps protect it against predators. A better camouflaged zebra has more chance of escaping a lion and therefore living long enough to breed. Thus, in an original population of unstriped zebra, if a striped form should appear, it would tend to be preserved across generations because the stripes have greater survival value. To use Darwin's terms, *natural selection* has favoured the selection and preservation of stripes.

Evolution by natural selection based on random variations within a species is a keystone of present-day biological thinking, though there are still very deep problems associated with trying to use evolutionary models to explain all aspects of the biological world. The point here is that, once again, biologists cannot understand the present form and function of living organisms simply by examining their present cell biology and biochemistry (reductionism) or their ethology or ecology (holism). It is necessary to understand their history (development and evolution) as well. The relationships between these different types of biological explanations are summarised in Figure 1.3.

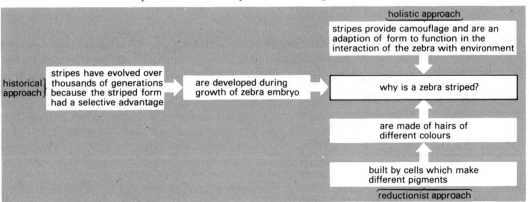

Figure 1.3 Explanations of biological form: reductionist, holistic and historical approaches.

In this book we attempt to consider the health and disease of humans from all these biological perspectives. Faced with the choice as to where to begin, we decided to commence with a reductionist view and discuss first the molecules and cells that make up the tissues and organs of the body (Chapters 2 and 3) and then turn to the development of an individual from its embryonic to its adult form (Chapter 4). In the next six chapters (Chapters 5–10) we consider the physiology of body systems which serve particular functions as diverse as excretion and conscious thought. From the discussion of normal functioning, we turn to the abnormal or diseased state. In Chapter 11 we consider why, in evolutionary terms, humans and other organisms suffer from disease. Chapters 12–16 are organised on the basis of the main categories of causes of disease, such as the physical environment and an individual's genes.

This century, and the period since the 1950s in particular, has seen an explosion of biological knowledge. Many problems seemed insoluble half a century ago. The detailed molecular structure of the complex substances which comprise living organisms, the mechanisms of the transfer of genetic information between parents and offspring, the coordinated workings of nerve and muscle are now understood with a degree of precision that, until quite recently, would have seemed inconceivable. With this new knowledge, society appears to be on the threshold of being able to intervene in the very processes of life, death and reproduction themselves. This new knowledge has come about by the application of scientific methods, coupled with the power of the new technologies that have made possible high magnifications and vast centrifugal fields, and the new methods that can precisely isolate a single molecule among millions of others—an achievement equivalent to being able to swat a fly blindfold in the dome of St Paul's Cathedral. The story of modern biology is one of the most exciting scientific adventures of all time. Yet the value of this vast array of facts about health and disease which biology can now offer can be judged only by considering the impact it has had on preventing, alleviating or curing disease. A brief assessment of the effectiveness of biomedicine forms the final chapter of the book.

Three final points before moving on. We have chosen to tell our biological story in the chapters that follow as a thoroughly modern and up-to-date set of descriptions and theories. We have not told you where the theories came from, how the facts were discovered, what the key experiments were. That is deliberate, but you should never forget that biology is an experimental science. The facts we describe come out of experiments: the theories have the same 'provisional' status that all scientific theories have.

The second point is that science works extensively by metaphor. The properties of one type of object in the universe are described in terms of another type that is better understood. From the earliest history of biology, the metaphors used in the study of living creatures were those of human-made machines. The energy processes of the cell are described as if the cell were a coal-fired power station, the heart as if it were a pump and the behaviour of the brain as if it were a super-computer. Chapters 5–10, on human physiology, are framed as if the organism could be understood as a series of *mechanical systems*, controlled, regulated by feedback circuits and adaptive to environmental change. You must be clear that these descriptions are metaphorical. They are *ways of seeing* the biology of organisms and, as such, conform to a current fashion. At present they seem to be useful, powerful ways of describing something of the reality of the life of the organism. They are, ultimately, only metaphors derived from the limits of human experience. The future may change our fashion in metaphors.

Thirdly, in the chapters that follow we shall often talk about 'cells' or 'organs' performing this or that activity, *without specifying the type of creature from which they are derived*. Living creatures differ enormously in form—we can all tell a human from a rosebush or a mushroom or a chimpanzee — but one of the consequences of the common evolutionary origin of all the different types of life is that a lot of their biology is also held in common. Molecules with remarkably similar structures do similar jobs in widely differing types of living creature. Shown a cell under a microscope, a cell biologist is more easily able to describe it as being from a brain or a kidney than from a human or a rat. The uniformity of biological principles is fundamental to our understanding of biology. You may well wonder what the point of biology is if it can tell us that more than 95 per cent of human cellular and molecular properties are identical to those of a chimpanzee, whereas it cannot, yet, account easily for the differences that everyone can see between humans and chimpanzees—but that is a different issue!

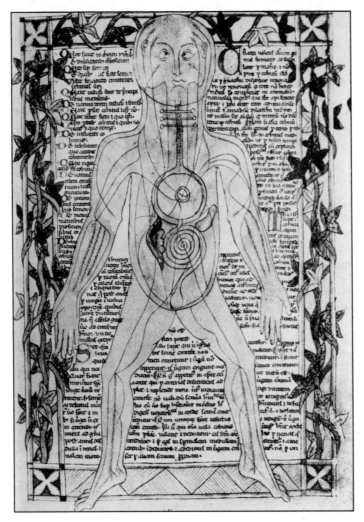

English thirteenth century miniature of the blood system and gut.

Objectives for Chapter 1

When you have studied this chapter, you should be able to:

1.1 List the characteristic properties of living organisms.

1.2 Explain what is meant by organisational hierarchy and the different approaches that biology adopts at each level.

Questions for Chapter 1

1 (*Objective 1.1*) Which of the following human properties are common to *all* living organisms?

   (a) conscious thought

   (b) growth

   (c) the repair of damage caused by injuries

   (d) the ability to feel pain

   (e) the ability to alter the external environment

2 (*Objective 1.2*) Summarise the types of questions that the following sciences would ask about the life of an adult cat: ecology, cell biology, biochemistry, developmental biology, and evolutionary biology.

2
The building blocks of life

In Chapter 1 we identified one of the key questions that biologists ask of living organisms: what are they made of? One of the extraordinary things about the enormous diversity of living organisms is that they all turn out to be built along fundamentally similar lines. If you were to take an animal or a plant or a fungus, and look at it under a microscope using higher and higher magnification, you would eventually discover that it was made up of a series of box-like repeating structures. All living organisms are built up of these unit structures, called *cells*. Some microorganisms consist of only a single cell, but most organisms are built up of millions of them. What are the cells themselves made of? How are they formed? How do they survive? When a living creature grows or reproduces, it does so by increasing the number of cells it contains, by a process of cell division. How does this replication take place? In this chapter we shall try to answer some of these questions. To do so, we must begin not with cells, but with the basic building blocks of which they themselves are composed: molecules of specific chemical substances. Let us start then with the chemical basis of life.

The chemical basis of life
All living matter is made up of chemical elements such as carbon, hydrogen, oxygen, etc. There are about 100 different elements, and all the myriad substances in nature, from deepest space to the inside of the human brain, are made up, ultimately, of these elements. The smallest possible particles of an element are termed *atoms*, and the smallest naturally occurring form of each substance is the *molecule*, made up from atoms of one or more chemical elements.

The elements in a specific molecule are combined in defined proportions. For instance, a molecule of water consists of two atoms of hydrogen (represented by the symbol H) and one of oxygen (represented by O). The 'chemical formula' for water is therefore written H_2O. The molecule of table sugar (sucrose) is more complex. Sucrose contains only one more element than water, carbon (C), but it has many more atoms, arranged in specific relationship to one another in space. Its formula is $C_{12}H_{22}O_{11}$.

Molecules that contain only a few atoms (such as water) are very tiny. To give you some idea of scale: if a protein (a relatively large molecule) were enlarged to the size of a football, a real football enlarged by the same amount would be the size of the Earth!

Living creatures contain many different types of small molecules, from common ones like ordinary salt (sodium chloride) to trace amounts of rare metals, but the characteristic molecules of life are large and complex. They are predominantly composed of three elements— carbon, oxygen and hydrogen— with lesser amounts of nitrogen. These chemicals are termed organic to distinguish them from the inorganic chemicals such as the salts and the metals, which are much more abundant in rocks and other non-living materials and which are made up of many additional elements.

Look at Table 2.1. This shows the composition of the human body in terms of the elements it contains. The most abundant is oxygen, followed by carbon and hydrogen.

☐ What do you think is the most abundant molecule in the body which accounts for much of that oxygen and hydrogen?

■ Water, which comprises 60 per cent by weight of the body. It is necessary for most of the chemical reactions that go on within the body.

Most of the solid mass of the body is composed of large and complex organic molecules containing varying amounts of carbon, hydrogen, oxygen and nitrogen. These large molecules fall into four general classes, three of which should be familiar to anyone who has ever read a diet book, because they are also the components of human food.

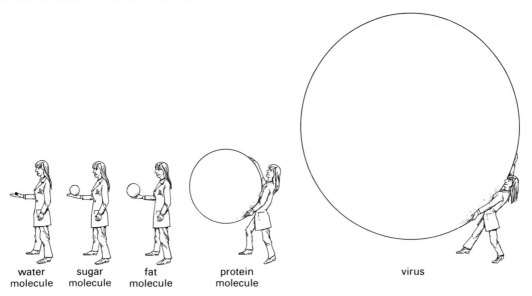

water molecule   sugar molecule   fat molecule   protein molecule   virus

Figure 2.1 The relative sizes of some molecules found in living creatures compared to a virus.

Table 2.1 Elements found in the human body

Element	Percentage of body weight
oxygen	65
carbon	18
hydrogen	10
nitrogen	3
calcium	2
phosphorus	1.1
potassium	0.35
sulphur	0.25
sodium	0.15
chlorine	0.15
Magnesium, iron, manganese, copper, iodine, cobalt, zinc	traces

    ☐   List these classes.

    ■   Carbohydrates, proteins, fats and nucleic acids. (You may not have got the nucleic acids.)

Each of these terms refers to a class of large molecules which share certain chemical characteristics. Carbohydrates and fats contain only the elements carbon, hydrogen and oxygen. Proteins and nucleic acids also contain nitrogen and nucleic acids also contain phosphorus. Although the chemical formulae of these compounds are very complex, their structures at least, are relatively straightforward. Carbohydrates, proteins and nucleic acids are all built up by joining together repeating units — chains of tens or hundreds of smaller, simpler molecules. You could think of them as strings of poppet beads. Their structures and compositions fit them for specific roles within the body.

Carbohydrates consist of strings of sugar molecules, for instance glucose ($C_6H_{12}O_6$), joined end to end. In animals, carbohydrates, such as glycogen, act as energy stores. In plants they can also act as structural molecules, giving plants their characteristic shapes and looks. For example, cellulose is found in the structure of all plants.

Fats (also known as lipids) store energy, provide body insulation and have a structural role in cell membranes. They are built up in a slightly more complex way, from several different types of subunit strung together, but it is not necessary to consider the details here.

Proteins, such as insulin and the blood protein haemoglobin, are chains made up of another type of basic building unit, the amino acids. Amino acids themselves are combinations of carbon, hydrogen, oxygen and nitrogen.

There are twenty different types of amino acids in living organisms and the order in which these different types are joined together determines the properties of the protein. Some, like haemoglobin and insulin, are soluble in body fluids, others are fibrous and insoluble, like keratin, the protein which constitutes hair, and collagen, one of the principal proteins of skin.

A very large number of different protein molecules can be produced by different permutations of the twenty amino acids found in living organisms. Thus in the cells of the human body there are probably about 100 000 different types of protein. One particular group of proteins, called *enzymes*, is of special importance. Their function, as you will see, is to catalyse (speed up) the chemical reactions going on continuously within the body.

Nucleic acids are also built up of units — called nucleotides. They have a specific function in the body about which we shall have a lot to say shortly. Briefly, they are responsible for transmitting 'information' from generation to generation during reproduction and they also control protein synthesis and cell division.

Living processes

The complex molecules of which the cell itself is composed do not occur in non-living nature. The living cell needs to make these molecules from other, simpler molecules. It achieves this by way of a complex series of chemical reactions. There are two central needs: energy and organisation.

Where do cells get energy from? Most cells and most organisms derive the energy they need by taking in substances (food) from their surroundings and chemically converting them to other substances, 'releasing' energy in the process.

This is a gross simplification. The 'energy' released is not generally in a form with which we are familiar — heat, light, or sound—nor is it, strictly speaking, released. Much of the energy is technically called free energy, though this is a most misleading term. It is 'free' only in one particular sense — free to 'drive' other processes, most often other chemical reactions that require energy.

So a cell may use energy to take up substances from its surroundings, by what is often an *energy-consuming* process. Some of these substances it then converts into others by means of chemical reactions. In contrast, some of these reactions are *energy-producing*. This energy is, in turn, used in other, energy-consuming, reactions either to manufacture required substances or to do work as when a muscle cell contracts, enabling an animal to move. It is important to remember that overall *every* cell must be able to 'balance its books'—the *energy used* by a cell in energy-consuming processes cannot exceed the *energy produced* by energy-producing reactions in the cell.

Energy is important, but it is not the sole product or resource of all this activity. A cell can be compared to a chemical factory: it takes in raw materials and fashions them into other materials. Some of the products, just like factory-produced goods, are exported from the cell, others are retained within the cell. Some become part of the fabric of the cell itself, for, just as in any factory, the cell 'machinery' wears out and must be replaced—continuous maintenance ensures continued function. Unlike most factories, the cell makes and replaces its own worn-out bits and pieces from within, making them from the raw

materials it obtains from its surroundings. *Replacement and repair of worn-out bits and pieces is a feature common to all cells.*

How are all these multitudinous chemical reactions organised within the cells? There must be some way of controlling and coordinating them. It turns out that, far from being a random hotch-potch of disorganised chemicals, cells are compartmentalised both structurally and chemically. The structural compartments of a cell are indicated in Figure 2.2, which shows a 'bacon-slice' section through a typical animal cell. Cells from different organisms and different organs and tissues within any one organism show various specialisations, but the general features in the figure are seen in most cells. Remember that the figure shows only a slice. The actual cell is in three dimensions.

We shall have more to say about the internal structure of a cell shortly. For the moment you should notice that the cell is surrounded by a thin *cell membrane*, which holds it all together, and that at its centre there is a *nucleus* surrounded by its own membrane. The rest of the cell is filled with a jelly-like solution of proteins and other chemicals, called *cytoplasm*, within which are a variety of structures. Remember that the 'structures' you are seeing are built from proteins, lipids, nucleic acids and carbohydrates, the 'giant molecules of life'.

Bearing this internal structure in mind, let us return to the chemical reactions that go on within the cells. Chemical reactions occur in linked series, called metabolic pathways (*metabolism* is the name given to the cell's chemical activity overall). A cell does its chemistry by means of many such pathways, involving thousands of individual chemical reactions. These pathways sometimes interlink and feed

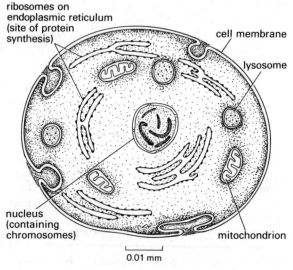

Figure 2.2 Cross-section of a typical animal cell.

(a)

(b)

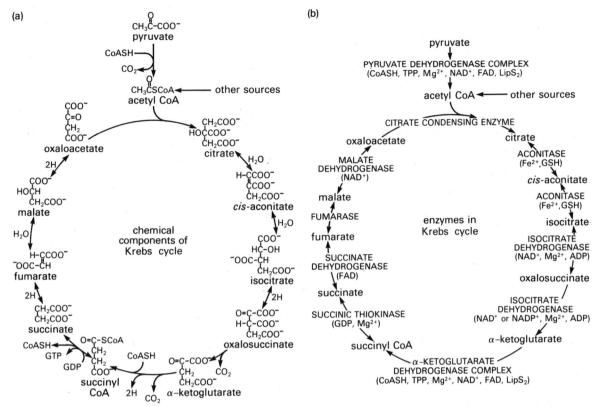

Figure 2.3 Example of a metabolic pathway (Krebs cycle) showing (a) the chemical components and (b) the enzymes involved.

into one another. Look, for example, at Figure 2.3, which shows a metabolic pathway called the Krebs cycle, named after its discoverer, Hans Krebs, who worked in Sheffield in the 1930s. (Krebs cycle is often now referred to as the TCA cycle.) You are not expected to remember any of the details of this cycle. It is shown merely to give you an idea of the complexity of chemical reactions in cells!

Krebs cycle is a set of important energy transfer reactions, which are part of the sequence by which the cells in animals, plants and many microorganisms break down sugars like glucose ($C_6H_{12}O_6$) to yield the 'waste' gas, carbon dioxide (CO_2), and water (H_2O). During this process, the energy released by the controlled 'burning' of glucose is used to produce molecules that form the starting point for the manufacture of proteins, nucleic acids and so forth.

The reactions of metabolic pathways are linked in this way for one major reason. They are 'step-by-step' reactions: each chemical in the sequence of the pathway differs only slightly from the next in the series. We shall mention just two of the myriad of sequences: that by which glucose is burned to carbon dioxide and water, and that by which amino acids are turned into proteins.

☐ Why do you think that chemical reactions are carried out step by step in the cell?
■ Small changes are easier for the cell to carry out in a controlled way. Not every chemical compound can be transformed into a second compound in a single step, even if they both contain the same elements. However, major transformations can be achieved through a large number of small changes.

These chemical changes are catalysed by enzymes. In the absence of enzymes, the reactions would occur only slowly or only at high temperatures. Enzymes are thus the cell's main resource in carrying out its very complex chemistry at reasonable rates. In general, each enzyme is tailor-made for only one reaction within the cell.

The study of enzymes gives valuable information about the control of cell processes and, at a higher level, about the working of the body. Much of this information has been gleaned from analysis of abnormal functioning. If a particular enzyme is defective or absent, the effect on the body can be extreme.

☐ Why is this so?

■ Because the particular reaction catalysed by that enzyme cannot take place, and consequently none of the reactions later in the sequence can occur.

For example, there is a condition known as galactosaemia in which infants are born without the enzyme that catalyses the production of the sugar glucose from another sugar, galactose. The absence of this enzyme leads to a rapid build-up of galactose in the body and death results after only a few weeks. However, a galactose-free diet solves this problem and, by avoiding cow's milk, from which the galactose is derived, it is possible for people who lack the enzyme to lead a normal life.

The importance of enzymes to body function also helps to explain how some drugs work. Drugs can be designed to interfere with the enzymes present in particular types of cell. For instance, the class of drugs called sulphonamides has been used extensively in controlling bacterial infections. The active agent stops one of the enzyme-catalysed steps in the production of the bacteria's nucleic acid. It therefore prevents the bacteria from growing. Humans, however, make nucleic acid by a different process, one which is not affected by sulphonamides and thus the drugs provide a safe and reasonably effective treatment.

Energy in cells

Each day the average human uses up enough energy to heat about 50 litres of cold water to boiling point. Where does the energy used in biological processes come from? As we have hinted, 'biological energy' is obtained in a very similar way to that in which heat and light energy are obtained from a fuel like coal. Coal is mainly made up of carbon and when carbon reacts with oxygen, energy in the form of heat is given off. This reaction is known as *burning* the coal. In the body, fats and carbohydrates (such as glucose) react with oxygen, and a steady flow of heat and other forms of energy result from *oxidation*, as the body's 'burning' processes are referred to.

How does oxygen react with fats and carbohydrates? The first task is to get oxygen into the body by breathing, a process called *respiration*. Oxygen in the air is taken into the lungs and carried by the blood to the cells of the body, where it participates in breaking down fats and carbohydrates in small enzyme-controlled steps. Normally, it is carbohydrates which are 'burned'; when fats are utilised to release energy, they are first converted into chemicals related to sugars and are then fed into the metabolic pathway that breaks down glucose. Strenuous exercise or the eating of a diet low in sugars causes the withdrawal of fats from tissues to keep the body supplied with energy. In extreme malnutrition, proteins can also be converted into substances which lie on the metabolic pathway of sugar

oxidation and hence used to make energy, but this is very much a 'last resort'.

During the oxidation of sugars in the cell, glucose is normally 'burned' completely to carbon dioxide and water, but this requires oxygen. In the absence of oxygen, glucose cannot be oxidised completely to carbon dioxide and water. It can, however, be converted to intermediate compounds. Anyone who has ever done home brewing will be familiar with at least one example of this type of intermediate.

☐ Can you name it?

■ In the absence of air, yeast converts sugars (such as glucose or sucrose) to alcohol.

This process does not yield as much energy for the cell as completely burning glucose to carbon dioxide, but it is possible for human body cells to respire anaerobically (without oxygen) for a short time if necessary. For example, during strenuous exercise not enough oxygen can get to the muscles to burn completely all the glucose and another chemical, lactic acid, is produced. The build up of this acid in muscle tissue is responsible for the sensation of cramp.

The energy released when glucose is burned is not allowed to dissipate, of course, but is stored in the form of a highly reactive chemical that can take part in many chemical reactions and in a vast number of other cellular processes. This substance, adenosine triphosphate, or ATP for short, is known as an *energy transfer molecule*. ATP has three atoms of phosphorus (in the form of phosphate) in its structure. It is made from a related molecule, adenosine diphosphate (ADP). The central role of ATP in the cell's 'energy economy' was discovered by the biochemist Fritz Lipmann, in the USA, during the 1940s. It takes a lot of energy to add the third phosphate group, converting ADP to ATP. The energy generated in cells can therefore be 'stored'. As and when necessary, an enzymic reaction splits off the third phosphate group from ATP; ADP is then recovered and the 'stored' energy is released.

Energy-requiring + ATP → Product of + ADP
reaction   reaction

ADP is rather like a discharged battery. It must be 'charged' to ATP again before being available for another energy-requiring reaction. The recharging process is possible because the ADP→ATP change is a by-product of other chemical reactions in the cell, such as the oxidation of glucose.

Protein synthesis and cell division

The enzymes involved in the production of ATP are made of protein, as is most of the cell. How are proteins themselves made? As we have said, proteins are molecules that consist of a sequence of amino acids, perhaps hundreds or thousands long. Every species of living organism has its own unique and characteristic set of proteins. Because they carry out very precise tasks in the cells, it is essential that they are themselves made accurately. A 'mistake' in joining up the amino acids, so that just one is in the wrong place, can result in an enzyme or other protein which simply cannot do its job. It is therefore — literally — vital that proteins are built accurately, and that this process can be repeated millions of times during the life of the cell. For these reasons, rather than build up each protein step by step as required, a mechanism for 'pressing out' the protein molecules on pre-formed templates has evolved. These templates are themselves built from the nucleic acids.

In humans, as in most living organisms, the nucleic acid templates are made of *deoxyribonucleic acid* (DNA). The structure of DNA—that is, how its molecular subunits, the nucleotides, are arranged in sequence to form the nucleic acid chain—was worked out in 1953 by Francis Crick, James Watson and Maurice Wilkins from the data of Rosalind Franklin. However, the existence of DNA had been known for many years before then, in fact, ever since the development of microscopes powerful enough to look inside cells.

When a cell has grown enough, it divides to give two daughter cells. Cell division, or *mitosis*, involves a complex series of events divided into phases, as shown in Figure 2.4. Filaments, or *chromosomes*, which consist of DNA wrapped in protein, appear in the cell during the early prophase. The number of chromosomes in a cell is always the same for individuals of a species: for example, 46 in humans; 14 in peas. The chromosomes condense into short, sausage shapes and the cell makes a duplicate copy of each chromosome (late prophase). They assemble in the centre of the dividing cell (metaphase) and then one copy of each chromosome moves to each end (anaphase). This process results in two identical sets of chromosomes, one at each end of the cell (telophase). The cell then becomes pinched in the middle, rather like a balloon with a string tied round its centre. Finally, the two halves separate. The result is two daughter cells, each with a single set of chromosomes whose DNA carries the information to repeat the entire process of growth and division.

The process of chromosome duplication is enormously complex and its elucidation has been a major triumph of modern biology. It hinges on the precise and elegant structure of DNA. As we have said, DNA is a complex chemical made of repeating units called nucleotides. Each nucleotide itself consists of a sugar molecule, phosphorus (in the form of phosphate) and one of four different nitrogen-containing substances called bases — adenine (A), guanine (G), cytosine (C), and thymine (T). (There is no need to remember these names.) These four substances are the basis of the so-called 'genetic code'. The sugar and

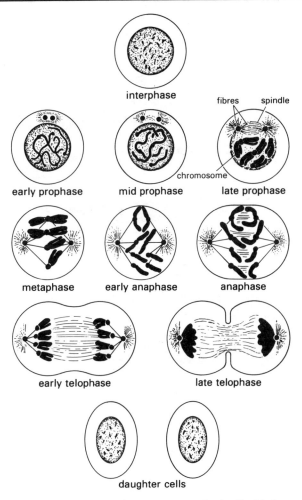

Figure 2.4 The stages of mitosis in an animal cell with four chromosomes.

phosphate molecules are arranged alternately in two long chains that spiral around each other, forming the famous double helix (Figure 2.5). This makes the 'backbone' of the DNA molecule. Across the middle, like the rungs of a ladder, the strands are joined by pairs of bases, which are attached to the sugars in the backbone chains. The most important property of the bases is that they have *complementary* shapes and can therefore pair up only in specific combinations across the helix—A with T, and C with G. Each pair fits together like a lock and key. This interlocking of shapes maintains the structure of the helix and is the key to the unique way in which perfect copies of the DNA can be made during cell division, and the way in which it can act as a template for the manufacture of specific proteins.

The way in which the DNA molecule is copied is beautifully simple, given the precise fit between the pairs of

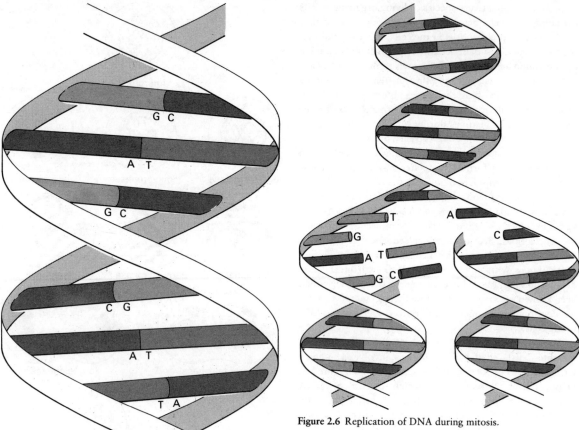

Figure 2.6 Replication of DNA during mitosis.

Figure 2.5 Structure of DNA, showing the double helix and base pairing.

bases. The process is shown in Figure 2.6. Starting at one end of the DNA chain, the pairs of bases are pulled apart by an enzyme-catalysed reaction that uses a lot of energy. The double helix 'unzips' and the two backbone chains separate. Supplies of bases, sugars and phosphates are brought in by different enzymes and the new bases pair off specifically, according to shape, with those on the old strand, now exposed and held apart by enzymes. Thus, two new DNA molecules are built up, each containing one old strand and one new, and *they are identical* by virtue of the complementary base pairing. New cells therefore contain the same genetic information as their 'parent' cell.

The process just described shows how it is possible to make identical copies of molecules of DNA during cell division, but DNA is also used to provide templates for making protein molecules. Protein molecules, as you know, are made up of sequences of any of the twenty different amino acids.

☐ What are the different 'units' of the DNA chain and how many are there?
■ They are nucleotides containing one of the four bases, A, G, C and T.
☐ How can the four bases code for twenty different amino acids?
■ Obviously, if there were twenty different bases, it would be easy, but since there are only four, various unique permutations stand for each of the amino acids.

If that is not clear, think about how, in morse code, permutations of only two symbols—a dot and a dash— can code for the twenty-six letters of the alphabet. If the four bases were arranged in groups of two, you can easily check that there are sixteen unique ways of arranging them (4 × 4). Since there are twenty amino acids, to be sure of unique codes for them all, it is necessary to take groups of three bases (a triplet) as the 'code word' for each amino acid.

□ How many 'coding possibilities' would this give?

■ 64(4 × 4 × 4).

Research has shown that the genetic code is indeed a *triplet code*, each triplet specifying one amino acid. As there are 64 possible permutations of three bases, some of the twenty amino acids have more than one triplet code. To make a protein such as haemoglobin, 300 amino acids long, requires a piece of DNA code 900 bases long. To make all of the 100 000 + proteins of the human body would require the equivalent of a length of DNA millions of bases long.

In humans, the DNA is distributed between the 23 pairs of chromosomes; each chromosome contains the DNA necessary to make quite a large number of the proteins of the body. Specific proteins are made only according to requirement. Synthesis is therefore not a question of starting at one end of a chromosome and making protein according to the DNA sequence all the way along to the other end. It is a more piecemeal affair. The DNA is, in fact, compartmentalised into sections called *genes*. Between the genes is so-called 'spacer DNA' whose sequence is not used as a code for protein. Indeed, the total DNA of any human cell (and remember that all body cells in any organism contain an identical set of chromosomes with identical sequences of DNA along them) amounts to some 1 000 million bases. There are therefore more than a million times more DNA than is required to code for all the proteins. The function of this extra DNA, which is probably concerned with helping regulate when, and in which type of cell, particular proteins are made, is still unclear and is the subject of one of the most active current areas of research in molecular biology. For the moment you need merely note that proteins are made according to the sequence coded in the DNA of genes arranged along the chromosome and that some of the triplet base 'code words' do *not* code for amino acids, but function as 'stop' and 'start' messages for the coding process. It is rather like beginning a sentence with a capital letter and ending with a full stop. The process whereby the cellular machinery 'reads' the DNA code and turns it into protein is called *translation*, for reasons which are, perhaps, obvious.

Proteins, however, are not made directly from DNA. The protein synthetic machinery is in the main body of the cell whereas the chromosomes are in the nucleus. Therefore, it is necessary to make a 'messenger' molecule that can take a transcribed copy of the genetic code from the nucleus into the cytoplasm. Only there can the information be decoded and translated into protein. The intermediary molecule is called messenger ribonucleic acid (mRNA). RNA is very similar to DNA, but the important difference is that RNA is single-stranded: that is, it has only one sugar–phosphate strand. Its bases can pair with only one strand of the DNA. mRNA is made in a similar way to

a new strand of DNA. The double DNA helix is unzipped by enzymes, and a molecule of RNA is built up, its base sequence being determined by the DNA on which it is built. This process is known as *transcription*. When the enzymes synthesising the mRNA reach a 'full stop' code word, the message is complete. The mRNA molecule is detached from the DNA strand, which can then zip up again with its complementary, non-coding strand. The mRNA can now pass out of the nucleus and into the cell body. Here it is decoded by structures called ribosomes that 'read' the triplet sequence of the mRNA. With the help of more enzymes, amino acids are joined together in the order specified by the code to form a complete protein. The whole process of transcription from DNA through RNA to protein is shown in Plate II.

This then is the famous DNA → RNA → protein sequence that makes up the cornerstone of modern molecular biology. The story of its discovery over the past thirty years has made headlines and won Nobel prizes. It is undoubtedly not only elegant and exciting, but also—by way of the new techniques of biotechnology and genetic engineering — providing ways of tailor-making specific proteins. DNA codes for some medically and industrially useful proteins have been introduced ('spliced') into bacterial cells which make the proteins in bulk.

There are still a lot of unanswered questions in the DNA story: what all the 'extra' DNA is for; how it is controlled and regulated; how we can explain a situation in which all body cells contain the same DNA, yet make many different types of protein, depending on the specialised function that the cell has in the body. For example, a skin cell looks different from a kidney, muscle or brain cell. These differences depend on the different proteins the cells contain, not their DNA, which is identical. Therefore, there have to be ways of 'switching on and off' the genes which make these proteins. The control of what is referred to as *gene expression* is again far from fully understood. What is certain is that it must involve 'feedback' from the rest of the cell to the DNA. The relationship between DNA and protein can be represented diagrammatically as follows:

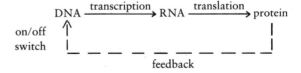

The structure of the cell

As you have seen, cells are composed of chemicals, of which many at any one time are involved in reactions that provide the energy for the cell's activities, for building new components such as proteins and for transmitting the information necessary for cell division to occur. Within the cells these chemical reactions are compartmentalised inside microscopic structures, known as organelles, such as the nucleus. Remember, these organelles themselves are also made from chemicals — they consist of lipids and proteins.

Look again at Figure 2.2, which shows the general features seen in most cells. You need to envisage this thin section in three dimensions. The main points to notice are:

1 Most cell structures are bounded by membranes.

2 The DNA is contained in the nucleus which, in turn, is surrounded by the nuclear membrane.

3 There are also fat sausage-shaped structures called mitochondria. Internal membranes (see Figure 2.7) increase the available surface area on which the energy-producing processes of the Krebs cycle occur.

4 A lot of the space outside the nucleus is taken up by an interconnecting series of tubes and pockets made up of membranes, the endoplasmic reticulum, in which many of the cell's chemical reactions occur. The endoplasmic reticulum appears as long oblong-shaped tubes in Figure 2.2 because of the way the section through the cell has been cut.

5 Some of the tubes described in (4) are directly concerned with storing and concentrating chemicals made for 'export' out of the cell: for example, enzymes that break down food in the intestines.

6 Attached to the tubes, or free in the cytoplasm, are small dot-like structures known as ribosomes, on which messenger RNA is 'read' and protein synthesised.

7 Also present outside the nucleus are round dark-coloured structures called lysosomes. These contain many of the enzymes concerned with the breakdown, inside the cell, of proteins, fats and so forth.

8 All these structures are bathed in a concentrated jelly-like solution, the cytoplasm, which fills the rest of the space inside the cell. The cytoplasm contains, in solution, many enzymes, glucose and other small molecules.

9 Finally, surrounding the cytoplasm, is the cell membrane.

Cell membranes are so important in the life of the cell that we must spend a little time considering how they are made and what they are for. A cell membrane must provide a boundary for the cell, but it must also allow the inward passage of 'friendly' substances and the outward passage of waste products. Cell membranes are made up of two classes of chemicals, proteins and lipids. The lipid layer forms the basic framework and the proteins are either embedded in this or closely associated with the inner or outer surfaces (Figure 2.8).

Two questions need to be answered: how is the membrane made and how does it work? The key to membrane structure lies in the form of the lipid layer, a structure composed of two single sheets. Each lipid molecule has two ends, one of which 'sticks' to other lipids. The other is repelled by the lipids and tends to attract molecules of water. Each of the sheets has its water-attracting part facing outwards, so the lipid-attracting parts are sandwiched in the centre of the membrane.

☐ Does the structure of the cell membrane give you any ideas as to how the membrane 'waterproofs' the cell?

■ The central 'sandwiched' parts of the lipids prevent water or water-soluble substances from crossing freely.

How then do materials cross a seemingly solid barrier? Some substances, such as water and water-soluble substances, pass through tiny 'holes', or pores, in the membrane. Substances that dissolve in fats can pass directly through the lipid layers. During the movement of protein through the membrane the protein molecules temporarily become incorporated in the structure of the membrane before being released to the other side.

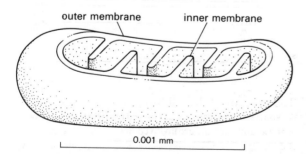

Figure 2.7 A mitochondrion.

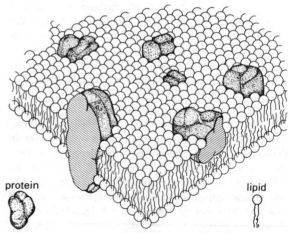

Figure 2.8 Structure of an animal cell membrane showing the arrangement of lipid and protein molecules.

The chemical reactions of cells occur in a fluid composed of water and dissolved substances. Within the cell, movement of fluids occurs by a process known as *diffusion*. This results in the movement of molecules from regions of high concentrations to those of low concentrations to 'equalise' matters. There is a balance between the fluid composition *inside* the cell and that of the environment immediately surrounding each cell. The cell maintains the correct internal chemical composition to sustain life, even when the external conditions differ from this optimum state. As you have seen, this process of balancing the internal with the external world is called homeostasis.

The fluid inside the cell contains organic chemicals, such as glucose, amino acids and ATP, but it also contains inorganic substances, such as sodium chloride (common salt). Other elements beside sodium form salts, compounds which, like sodium chloride, have a particular property when dissolved in water.

The molecule of sodium chloride consists of one atom of sodium, Na, and one atom of chlorine, Cl. When sodium chloride is dissolved in water, as it is in the cell, an interesting thing happens. It temporarily breaks down into its constituents, not in the form of free atoms (which would be highly reactive), but into electrically charged *ions*, thus:

$$NaCl \rightleftharpoons Na^+ + Cl^-$$

The sodium ion carries a positive charge and the chloride ion a negative charge. The single-headed arrows mean that the reaction is reversible. Similarly, the salt potassium chloride breaks down in water to give positive potassium ions (K^+) and negative chloride ions (Cl^-). Each of these solutions has the same number of positively charged sodium or potassium ions as there are negatively charged chloride ions.

In the different compartments of the human body the concentrations of ions varies. For example, the concentration of sodium ions is higher outside cells than within them, while the concentration of potassium ions is higher inside cells. Indeed, one function of the cell membrane is to maintain these different concentrations of ions.

☐ If the cell membrane allowed substances to diffuse freely into and out of the cell, what would you expect the composition of internal and external fluids to be?

■ They would be identical.

Actually, apart from the amount of water inside and outside cells (water tends to diffuse in and out freely), the concentrations of many chemicals differ between the two environments. In addition to potassium ions, amino acids and magnesium ions are more concentrated *inside* cells. Sodium ions, chloride ions, bicarbonate ions and glucose are all more concentrated *outside* cells.

These differences in concentration across the cell membrane are maintained by a mechanism called *active transport*. This uses energy to pump substances through the cell membrane, regardless of the concentration difference inside and outside the membrane, and allows the creation of appropriate fluid environments within cells.

Figure 2.9 shows the active transport of sodium and potassium ions across the cell membrane. Sodium inside the cell combines with a carrier molecule (a protein) at the inner surface of the cell membrane, and is then taken to the outer surface. Here the carrier molecule is modified (probably by an enzyme) so that it releases the sodium. It then combines with potassium, which it transports to the inner surface of the membrane, where energy is provided to split the potassium from the carrier.

The process which exports sodium, often referred to as the sodium pump, is important both for maintaining the internal composition of the cell environment and for a variety of functions in complex multicellular organisms. For instance, it is an essential part of the transmission of nerve impulses. The different concentrations of sodium ions and potassium ions on opposite sides of the cell membrane result in a difference in electrical charge across the membrane. This difference is essential to generate the 'bursts' of electrical activity which form nerve signals.

Membranes then are very important structural components of cells. The cell membrane separates the cell from the outside environment, and internal membranes, in particular those around the organelles, divide the cell into compartments, thus preventing it from becoming a disordered chemical 'soup'. Membranes also provide a matrix to which many of the enzymes that regulate chemical reactions in the cell are attached. The breakdown and build-up of many chemicals in the cell takes place in association with these internal membranes. In addition, the cell membrane has the ability to flow around small amounts

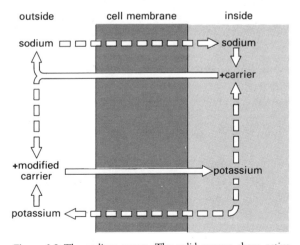

Figure 2.9 The sodium pump. The solid arrows show active transport of ions through a cell membrane. The broken arrows show diffusion from high to low concentration.

of substances present in the fluid outside the cell. The substance becomes completely enclosed by membrane, which is then 'pinched off'. Figure 2.2 shows this process in progress. This part of the membrane is then absorbed into the cell, together with its contents. It is called pinocytosis — literally 'cell drinking'. Pinocytosis is important because it is the only way that very large substances can be transported into most cells; whole protein molecules can be taken in, for example. In single-celled animals, pinocytosis is often the major feeding mechanism. For larger animals (such as humans) that have evolved a digestive system to break down food substances into their component molecules, pinocytosis is very much less important. A similar process, however, is used by some scavenging cells in the blood whose function is to clear out any intruding matter such as microorganisms. This process is known as *phagocytosis*—'cell eating'.

This concludes our discussion of the chemical and structural basis of life—a rapid and inevitably condensed tour through the major classes of molecules from which cells are made, the production of energy, the regulation of chemical reactions by enzymes, the unique ability of DNA to transmit information coded in its structure, the organelles of the cell and the role of cell membranes. If you have no previous experience of cell biology, you may be feeling a bit swamped by all this new information, but as you study this book you will find that we return again and again to concepts and terminology discussed here. This chapter is not only the 'structural basis of life' but also the basis of the book!

Objectives for Chapter 2

When you have studied this chapter, you should be able to:

2.1 List the main chemical constituents of living matter and distinguish between the major classes of large organic molecules.

2.2 Explain some of the basic processes of cells, including how energy is obtained; how DNA is replicated; and how RNA and proteins are made in cells.

2.3 Identify the principal subcellular components of the cell and their main functions.

Questions for Chapter 2

1 (*Objective 2.1*) Describe the difference in function in the human body of proteins, nucleic acids and carbohydrates.

2 (*Objective 2.2*) What are the principal steps in the cellular manufacture of a new molecule of an enzyme from its constituent subunits?

3 (*Objective 2.3*) Plate I shows a photograph of a liver cell. Identify the following structures: nucleus, mitochondria, lysosomes, endoplasmic reticulum, and cell membrane.

3
The cells of the body

You saw in the last chapter that cells are the basic 'unit' of biological function and that a number of characteristics are common to all cells. In this chapter we shall look at these characteristics in more detail, in both single-celled and multicelled creatures.

Single-celled organisms

The major single-celled organisms are bacteria, viruses and the single-celled animals and plants. As you will see in Chapter 14, although many of these organisms co-exist with humans in a mutually beneficial manner, some cause disease when they multiply within the human body.

Single-celled organisms display a range of different structures. Some, for instance yeasts and protozoa, are similar to the cells of all multicellular creatures in that they contain a nucleus with chromosomes within it. However, other single-celled organisms, such as bacteria, do not have a nucleus—as you can see by looking again at Figure 1.1. Their DNA is normally in the form of a long multicoiled helix and is not arranged into chromosomes as in other types of cell.

The single-celled organisms are in some ways less complex than the individual cells of multicelled creatures since they lack a number of cell constituents. On the other hand, because they have to exist independently and survive conditions which are more extreme than those encountered by cells inside the body of a multicellular organism, they often show a greater repertoire of biochemical versatility in the chemical reactions they can perform.

Viruses

Viruses occupy a special place on the borderline between life and non-life — which side of the line is chosen depends on which definition of life is being used. Viruses have no real independent existence: they do not possess the ability to move and consist of only a nucleic acid chain protected by a protein coat. Viruses replicate by entering other cells and 'persuading' the DNA of the cells they infect to make viral components instead of their normal output. The virus achieves this by first attaching itself to the surface of its *host* cell. The viral protein then digests away part of the cell membrane and the viral nucleic acid passes into the host cell, leaving the protein coat behind. The empty coat has no further function. The injected nucleic acid exerts control on the host's synthetic machinery and directs the production of daughter virus particles, each of which has a protein coat enclosing the viral nucleic acid. These particles leave the host cell (which is often killed in the process) and the cycle starts anew. Figure 3.1 shows adenoviruses and influenza viruses, both of which are involved in diseases of the upper respiratory tract.

Viruses are therefore 'genetic parasites' which take over and redirect the nucleic acid of the host cells they infect. Because the host cell does all the chemical synthesis required to make new viral particles, the virus itself does not need a complex synthetic machinery.

Bacteria

Bacteria are the simplest 'true' cells—a typical bacterium was shown in Figure 1.1.

☐ What is the major difference between bacteria and a typical animal cell such as that shown in Figure 2.2?
■ Bacteria have no nuclei or chromosomes and do not contain mitochondria.

Bacteria are able to divide very frequently. Under optimum conditions it is not impossible, beginning with a single bacterium, to produce tens of millions of bacteria in a day, each cell dividing once every 20 minutes. The cell membrane of many types of bacteria is surrounded by a tough outer wall made of cellulose to protect it from the often hostile environment in which the creature may find itself. In addition, many bacteria are capable of producing spores, 'life-support capsules' which can withstand and resist a variety of extremes, including drying out and chemicals, such as antibiotics, which are normally lethal.

Bacteria, although very small, are relatively large when

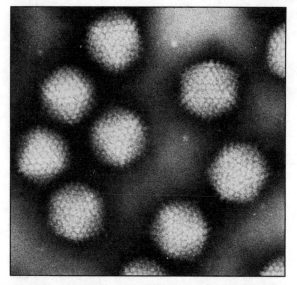

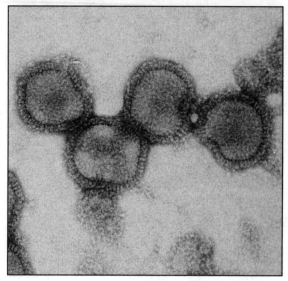

Figure 3.1 Particles of (a) adenovirus (magnification 110 000 times) and (b) influenza virus (magnification 145 000 times).

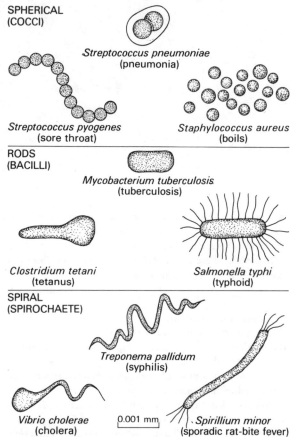

SPHERICAL
(COCCI)

Streptococcus pneumoniae
(pneumonia)

Streptococcus pyogenes
(sore throat)

Staphylococcus aureus
(boils)

RODS
(BACILLI)

Mycobacterium tuberculosis
(tuberculosis)

Clostridium tetani
(tetanus)

Salmonella typhi
(typhoid)

SPIRAL
(SPIROCHAETE)

Treponema pallidum
(syphilis)

Vibrio cholerae
(cholera)

0.001 mm

Spirillium minor
(sporadic rat-bite fever)

Figure 3.2 Variety of bacterial shapes, with examples of species and the diseases caused by them.

compared with viruses. A typical bacterium is only a few thousandths of a millimetre long and only a thousandth of the volume of a typical human cell. Bacteria have a variety of characteristic shapes, as shown in Figure 3.2.

Whereas viruses normally disrupt host cells of living creatures and cannot multiply without doing so, many bacteria feed on dead and decaying animal and plant matter. The danger to animals of some bacteria lies in the waste products that the bacteria produce. These pass out into the host's tissues and in some cases act as poisons (called toxins) which produce drastic effects. This is true of such diseases as cholera and tetanus.

Single-celled animals and plants

These organisms are thought to have evolved from bacteria-like single cells. The group is characterised by having larger cells than bacteria. Some are up to a tenth of a millimetre long and just visible to the naked eye, although many are only a few hundredths of a millimetre long. Single-celled creatures of this group are more mobile than bacteria and can often change shape. Reproduction occurs in a similar manner to that seen in bacteria.

An example of a single-celled animal (protozoan) is the *Amoeba*. *Amoeba* is the family name for a group of related creatures which live in watery fluids. Some may cause disease in the intestines of larger animals such as humans. Another example of a disease caused by a protozoan is malaria. Figure 3.3 shows the creatures described so far in this chapter, drawn to the same scale to give an idea of their relative sizes.

Multicellularity

There is a limit to the size of single cells, due partly to the distance over which chemical reactions can be controlled adequately. For example, diffusion of chemicals over long distances is too slow for efficient utilisation. There are also physical laws of structure which prevent cells from growing too large. The cell wall of a bacterium, a few thousandths of a millimetre in length, is perfectly adequate to maintain rigidity, but a cell the size of a human would tend to sag a little!

There is also another problem. Cells communicate with their external environment by transport across the cell membrane. As cells get larger, the ratio of the volume of the cell to the surface area of the cell membrane alters. Imagine two spherical cells, one with a radius twice that of the first. The surface area of the larger cell is *four* times that of the smaller cell (surface area increases as the *square* of the radius of a sphere). The volume of the larger cell, however, is *eight* times that of the smaller cell (volume increases as the *cube* of the radius). Molecular 'traffic' across the cell membrane will increase rapidly as the cell increases in size until, eventually, a point is reached where it is not possible to get sufficient oxygen in, and carbon dioxide out, for the cell to function. The evolutionary 'solution' to increasing size is multicellularity, which gives more scope for a wider range of features. The central question here is why multicellularity should be needed at all.

☐ The evolutionary answer to this question involves changes in the environment. Can you suggest an answer?

∎ Changes in the environment lead to the preferential survival of creatures best able to adapt to the new conditions. Scarcity of a particular food source, for example, would favour creatures which could switch to a different type of food. If vegetarians were physically unable to eat meat, then scarcity of plant material would lead to their demise. If a few of the vegetarians could 'adapt' to a meat-eating form, then they would survive.

In evolutionary terms, the more an organism is able to adapt, the better it is equipped to deal with adverse conditions. Thus individual varieties of bacteria are limited to particular environments. Some live in the human gut whereas others eat mineral oil, each variety being limited to its own 'niche' or habitat. Multicellularity, however, has allowed the evolution of resistant outer coverings, temperature control systems, sophisticated modes of locomotion and reproductive advantages, such as the ability to care for offspring. This increased ability to function in a wide range of environments allowed living organisms to spread, early in the history of life on Earth, to a huge array of habitats from the polar ice caps to the tropical jungles.

As multicellularity evolved, so the environments of cells changed. Only a small proportion of the cells of multicellular organisms interface directly with the outside world. Most are internal, surrounded by other cells. This means that their environment can be closely regulated and that they no longer need the full range of versatility and adaptability of single-celled organisms. Hence the cells of a multicellular organism can afford to be biochemically and physiologically specialised. This can be seen in even the simplest multicellular creatures, the fungi. Fungi consist mainly of threadlike branching tubes which contain cell material without complete separation between cells (Figure 3.4). Each fungus forms from a single spore, a single fungal

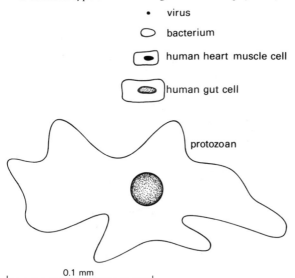

Figure 3.3 Relative sizes of viruses, bacteria and animal cells.

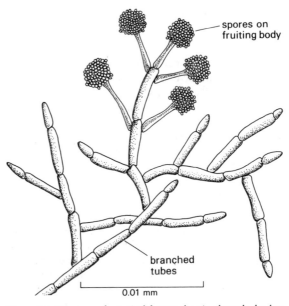

Figure 3.4 Structure of a typical fungus showing branched tubes and spore formation.

cell dispersed from an existing fungus. Two examples of fungi that affect human health are *Penicillium* (from which the drug penicillin is obtained) and *Aspergillus* (which can cause respiratory disease).

The animal body—cell differentiation

So far in this chapter we have looked at some of the advantages of multicellularity. The next task is to consider the layout of the animal body—how cells, tissues and organs are organised to work together. The concern here is with the structural relationship of body components, the questions of form. The detailed study of function must wait until Chapters 5–10.

The cells in multicellular organisms are not arranged haphazardly, but are grouped together according to *functional specialisation*. A living creature can be considered as a collection of cell populations, an hierarchically organised compendium (Figure 3.5). Similar cells are packed together into recognisable tissues such as muscle or bone. Several different tissues then make up organs such as the heart or liver. In turn, organs form a number of interrelated systems. Each 'system' is a collection of components organised to carry out a common task: for example, the respiratory system.

The respiratory system is composed of several organs such as the lungs and the windpipe (trachea). Each organ is made up of several different tissues. For example, the lung is composed of repeatedly branching air tubes that end in minute sac-like structures known as alveoli. The walls of these tubes contain several different tissues, such as muscle, which can alter the diameter of the tubes, and a special lining tissue known as epithelium.

We need to consider briefly how the various types of cells differ in terms of their internal functioning.

☐ Do you think that there would be any difference in the chemical reactions that go on in, say, a muscle cell and a cell in the intestines producing a digestive enzyme?

■ Yes, quite apart from the obvious difference that the intestinal cell synthesises an enzyme and needs to carry out the chemical reactions needed for this synthesis, the muscle cell has to produce supplies of energy to fuel the muscle contraction process.

☐ Which cell organelles are particularly involved in energy-producing reactions?

■ Mitochondria.

We might therefore expect that muscle cells would have a lot of mitochondria and this is the case. In contrast, cells that produce 'chemicals for export' (such as the enzyme-producing cells in the intestines) are rich in organelles that manufacture proteins.

Of course, the external appearance of the cells also differs. Muscle cells are long and spindle-shaped (Plate

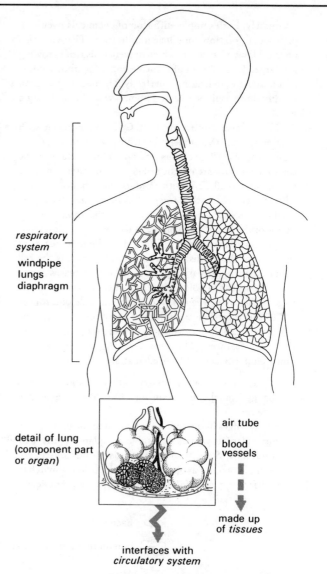

Figure 3.5 Organisational hierarchy of the respiratory system.

IIIa–c) whereas intestinal cells are squarer in appearance (Plate IIId). The spindle shape of muscle cells reflects their ability to elongate and contract.

Tissues of the body

Let us now consider the main types of tissues that make up the human body. In order to survive and act upon the world outside, the body of a multicellular organism needs to have a strong, flexible structure and certain cell types have as their prime function the maintenance of this structure.

☐ List all the cell types which you think might play this kind of role. (Think of the basic body tissues that may be involved.)

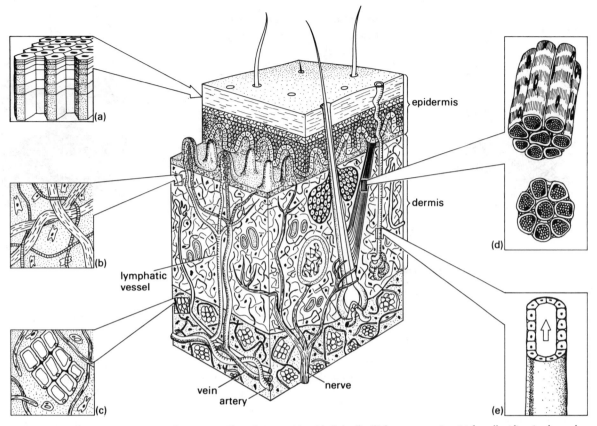

Figure 3.6 Structure of skin, showing the various cells and tissues: (a) epithelial cells; (b) loose connective; (c) fat cells; (d) striped muscle; (e) epithelia lining of duct.

■ You may have listed some or all of bone, cartilage, skin, muscle, tendons and fat.

Structural tissues are normally considered in three categories: bone and cartilage, which provide rigidity; connective tissue, which holds the internal organs in their correct stable locations and performs a variety of connective and supporting roles in the skeletal and muscular systems; and epithelial tissue, which covers the surface of organs, lines tubes and ducts and makes up the outer layer of the skin.

The way some of these tissues are arranged in the skin can be seen in Figure 3.6. Not all external surfaces of the body are covered by skin: for instance, the lining of the mouth and inner surface of the genitals. These structures are covered by mucous membrane which, unlike skin, lacks a horny layer and which secretes mucus, a slimy substance that keeps the membrane from drying out. Mucous membrane is more vulnerable than skin to injury or infection, the implication of which will be seen in Chapters 13 and 14.

In addition to structural tissues, there are several other tissue types: muscle tissues; nervous tissue, including brain cells as well as cells carrying nerve impulses round the body; reproductive tissue, which produces sperms and ova; and blood cells, of which there are several different types with different roles.

In the remaining part of this chapter we shall concentrate on three tissue types — bone and cartilage, connective tissue and muscle — and look at their different structures and functions in the body. The other tissue types will be discussed in later chapters.

Bone and cartilage
The body's scaffolding is provided by two related tissues, bone and cartilage. As you will see in the next chapter, during the body's early development, cells of the embryo become 'condensed' into blocks of cartilage cells that eventually give rise to bone cells, which themselves produce the structure that we recognise as bone. This process is gradual and is not complete by birth. It is not until late childhood that the replacement of cartilage by bone is completed. Even then many cartilaginous regions remain in

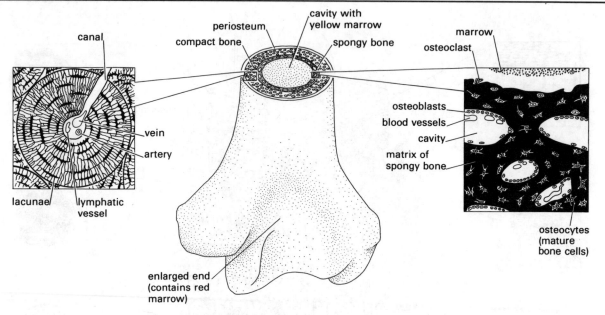

Figure 3.7 Structure of a long bone showing the surface covering (periosteum), the arrangement and content of compact and spongy bone and bone marrow. Note that hollow spaces (lacunae) occur in both types of bone.

the adult, for example in joints, because cartilage has the valuable properties of being softer and more resilient than bone.

Bones are not rigid non-living structures, but are comprised of networks of living cells embedded in a hard framework of protein and calcium- and phosphorus-containing substances (Figure 3.7 and Plate IIIe). Bone cells (osteoblasts) first produce fibres of a protein called collagen (see 'connective tissue' below). Once these fibres are present calcium salts are deposited to harden the matrix. Bone deposition is regulated partly by the amount of strain applied to the bone, so bone subjected to continuous loads grows strong whereas bones that are not used at all, such as the bones of a leg in a plaster cast, tend to waste away. Breaks in bones stimulate bone cells locally to produce large quantities of protein fibres in which calcium salts are then deposited to form new bone. As a result, the break is usually repaired within a few weeks.

There are more than 200 bones in the adult human body (Plate IV) and they may be divided into groups according to either their form (long or flat) or their function.

☐ What functions can you suggest for bones?
■ Support and protection for internal organs and support for lever systems (with muscles) in limbs.

An example of a long bone is the femur, running from the hip to the knee. It has a rounded shaft and is enlarged at the ends for articulation with adjacent bones. The shaft is hollow and is made up of compact bone enclosing a cavity

lined with a layer of spongy bone and filled with yellow marrow, which is the site of blood cell production in the body. Flat bones, such as the scapula (shoulder-blade), are made up of two compact bone plates between which is sandwiched a layer of spongy bone containing red marrow.

In healthy bone, new bone is continually replacing old and the construction and destruction processes are in equilibrium. Old bone is 'recycled' by the action of special types of bone cells (osteoclasts) which secrete powerful enzymes capable of digesting away bone material—the chemicals released by this process are then available for further bone production. The disease osteoporosis occurs when the equilibrium between bone production and bone digestion breaks down. This happens with ageing, so that bones become weak and brittle and break easily.

Joints enable articulation of one bone against another, and so allow movement. Some bones are closely joined in the adult by rigid connections called sutures: for example, the bones of the skull. These sutures do not form fully until about one year after birth.

☐ What advantage is there for a baby to be born without having developed a rigid skull?
■ It allows flexibility to facilitate passage through the mother's pelvis and room for the brain to grow after birth.

Most bones are joined by special joint structures (Figure 3.8). Where two bones touch, each has a covering of cartilage which is smooth, and softer than bone, giving the

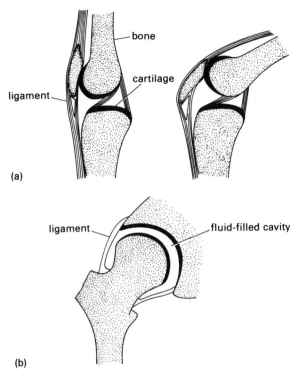

Figure 3.8 The structure of joints: (a) knee (side views) and (b) hip (front view).

ability to absorb jarring. Wear and tear on this cartilage may result in arthritis. The bone ends are enclosed in a watertight structure called the joint capsule. Some joints (those with freer movement) have an extra membrane inside the capsule, and this produces a joint-lubricating fluid. Tough bands of tissues called ligaments bind the whole joint together.

Specialised joints are found in the backbone (Figure 3.9). Here, thirty-three small bones called vertebrae are interconnected by joints and ligaments. With the exception of the first two vertebrae, which are modified to hold up the skull, and the last nine (the sacrum), which are fused together to some degree, all the vertebrae have the same general shape. This shape can be seen in Figure 3.10 and consists of an arch, which encloses the spinal cord, and the body of the vertebra, which is the main rigid part. Three projections are also present, one at the back and one at each side. The side projections articulate with the ribs (where these are present in the chest region) and help to support them. The top and bottom sides of each vertebra are modified to articulate with the vertebrae above and below, and also for the attachment of ligaments to support the back.

The joints between vertebrae consist of a flat, round intervertebral disc made of cartilage. Each disc has an outer fibrous covering and an inner jelly-like centre. The disc is

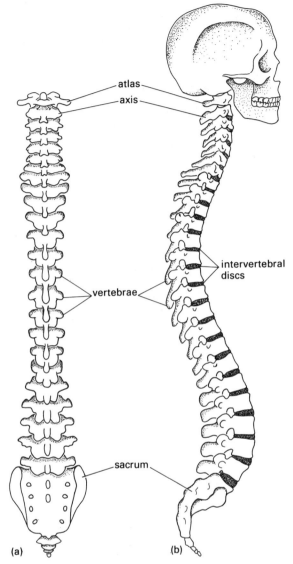

Figure 3.9 (a) Back and (b) side views of the spine showing the arrangement of the vertebrae and intervertebral discs.

attached to adjacent vertebrae by cartilage connections and ligaments between the vertebrae serve to strengthen the joint. Although each joint is capable only of limited movement, together the large number of joints in the spine gives considerable flexibility. The jelly-like centre of the intervertebral disc is an important component of this flexibility, as it can move backwards and forwards to relieve and redistribute local pressure as required (Figure 3.11). The popular description of some back pain as being due to a slipped disc is a misnomer: the disc does not actually slip out of place, but is ruptured or bent out of shape, and may press on the spinal cord, causing severe pain.

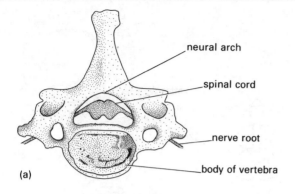

(a)

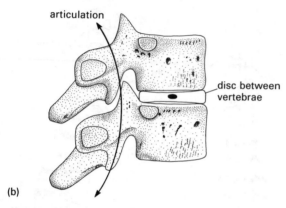

(b)

Figure 3.10 Typical vertebrae viewed (a) from above and (b) from the side.

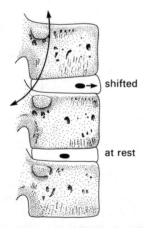

Figure 3.11 Articulation of the spine showing changes in the shape of the intervertebral discs (side view).

Connective tissue

Connective tissue is found throughout the body, and consists of protein fibres (collagen and elastin) and cells suspended in an amorphous, jelly-like substance. Collagen

is made by cells called fibroblasts, rather elongated cells which are spaced throughout the connective tissue (Figure 3.12).

Collagen has received some publicity lately due to the extravagant claims of some beauty creams either to contain or to 'condition' collagen in skin. The facts are that collagen changes its structure as it 'ages' and this change in structure is partially responsible for the 'wrinkles' in human skin as it grows older. To reverse the structural changes in collagen would require some drastic chemical treatments that would need the total digestion and reconstitution of the collagen fibres: clearly *not* the province of beauty therapy!

The several varieties of connective tissue serve to support and bind body structures together. *All* epithelial tissues, blood vessels, nerves and muscle cells are ensheathed with connective tissue fibres. The *density* of the connective tissue differs depending on its role. Loose connective tissue is characterised by a viscous matrix with, as its name suggests, a loose arrangement of fibres (Plate IIIf). It is found supporting the internal body organs and under the skin. Fatty (or adipose) tissue stores fat which can be broken down when reserves of energy are required. Fat tissue is also used for insulation, and is found under the skin (Figure 3.6c) and in a variety of other locations in the body. Dense connective tissue consists of masses of collagen fibres with fibroblasts interspersed among the fibres (Plate IIIg). It is 'tougher' than loose connective tissue (Figure 3.6b) and encapsulates cartilage and bone. It is also found in the dermis of the skin.

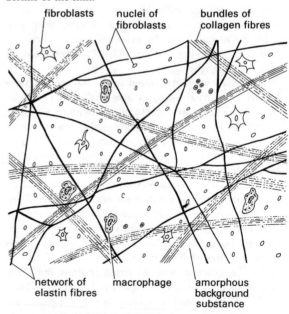

Figure 3.12 Structure of connective tissue showing cellular and fibrous contents.

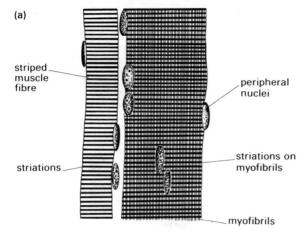

(a)

striped
muscle
fibre

peripheral
nuclei

striations

striations on
myofibrils

myofibrils

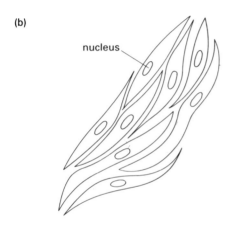

(b)

nucleus

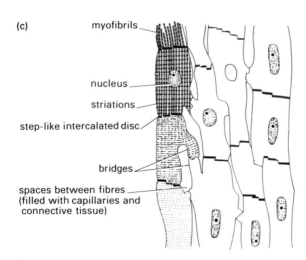

(c)

myofibrils

nucleus

striations

step-like intercalated disc

bridges

spaces between fibres
(filled with capillaries and
connective tissue)

Figure 3.13 Structure of (a) striped, (b) smooth and (c) cardiac muscle.

Muscle

Muscle cells are of three important and distinct types, all of which perform different tasks in the body. Striped or skeletal muscle is made up of large cylindrical cells which can be seen under the microscope to have a banded pattern (Figure 3.13a). This type of muscle is capable of strong contractions and serves to move bones. It can be contracted 'at will', as it is under the voluntary control of the conscious brain.

Smooth muscle (Figure 3.13b) consists of small, spindle-shaped cells and is found, for example, around the gut where it performs slow contractions over a long period. In general terms, smooth muscle occurs in the digestive, circulatory, respiratory and reproductive systems and is not normally under conscious control.

The third type of muscle, cardiac muscle (Figure 3.13c), has the property of being able to beat rhythmically without outside control. A strip of cardiac muscle will go on beating for some time even if it is removed from the body, provided it is bathed in a nutrient solution. It is thus well suited, as its name suggests, to make up the walls of the heart. Cardiac muscle fibres are branched and interconnected, so that electrical impulses can be transmitted through the whole heart.

All muscles work in more or less the same way (Figure 3.14). Muscle fibres contain two proteins (actin and

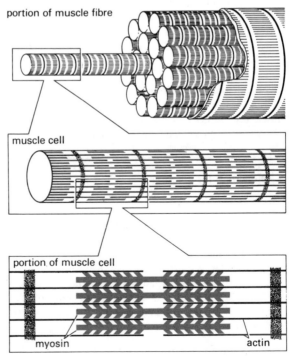

portion of muscle fibre

muscle cell

portion of muscle cell

myosin

actin

Figure 3.14 Structure of striped muscle showing the actin and myosin system for contraction.

myosin). These proteins fit together in the fibres and can move over each other in a ratchet-like mechanism, thus shortening the muscle fibre as they come together and lengthening it as they move apart. ATP is used up in the process of contraction.

In striped and smooth muscle, the signal for contraction is given by nerves which connect directly to the muscle cells. An electrical impulse passes down the nerve (nerve signal propagation will be discussed in Chapter 6), causing an electrical charge to be generated on the muscle fibre membrane. The charge induces a chemical change within the muscle cells, which is the signal for muscle contraction to occur. Contraction is an *active* process, but the muscles relax again *passively*. In some diseases (for example, tetanus) there is a sustained barrage of nerve impulses to the muscles and this produces a state of sustained contraction known as tetany (hence the familiar name for the disease —lockjaw).

Skeletal muscles are attached to the bones by means of tendons, which are tough connective tissue bundles. The muscle fibres themselves are bound together in bundles and the whole muscle is surrounded by a protective and constraining sheath. Skeletal muscles usually work in pairs, one to flex and the other to extend a body part. Thus one muscle contracts while the other is relaxed. For example, in the upper arm (Figure 3.15) the biceps and triceps muscles control movement. When the biceps muscle contracts, the triceps relaxes, and the arm bends. The arm is straightened when the biceps relaxes and the triceps contracts. Muscles which produce bending movements of a limb are called flexors and those which straighten the limb are called extensors.

Smooth muscle is controlled by nerve signals, but unlike skeletal muscle the nerves are connected to *bundles* of muscle cells rather than to individual cells. 'Pacemaker' cells are also present and these can initiate contraction

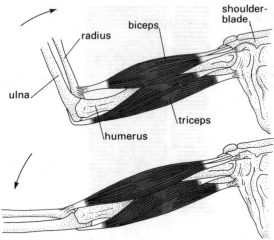

Figure 3.15 Triceps and biceps muscles of the upper arm showing an extensor and flexor in opposition.

waves such as, for example, the rhythmic waves of contraction which move food slowly down the gut. A further characteristic of smooth muscle is that it needs less energy to function than skeletal muscle.

In this chapter we have discussed the ways in which living forms can survive as single cells constantly confronting and co-existing with an external environment which may be alternately a beneficent source of food or hostile and threatening to the continued existence of the cell. We have also shown how one solution to this problem of maintaining an independent living existence has been the evolution of multicellularity, which allows the specialisation of cells in organs and tissues. We have still described cells only in their fully mature, functioning state. It is time to turn to another vital dimension of the mainstream of life, that of time. Over time, organisms grow, cells divide, reproduce and die. This is the subject of the next chapter.

Objectives for Chapter 3

When you have studied this chapter, you should be able to:

3.1 Discuss the main features of viruses, bacteria, and protozoa.

3.2 Summarise the advantages to animals of being multicellular, and explain the hierarchical relationship of cells, tissues, organs and systems in animal bodies.

3.3 Describe the features and function of the following tissue types: bone and cartilage; connective tissue; muscle.

Questions for Chapter 3

1 (*Objective 3.1*) Which of the following structures are possessed by (a) viruses, (b) bacteria, and (c) protozoa?

   (i) a nucleus   (iv) a cellulose cell membrane
  (ii) cytoplasm   (v) nucleic acid
 (iii) chromosomes

2 (*Objective 3.2*) Is the kidney an organ, a tissue or a system?

3 (*Objective 3.3*) Figure 3.6 shows the different tissues of which skin is composed. Where in skin do each of the following tissues occur and what functions do they fulfil?

   (i) epithelium   (ii) connective tissue   (iii) muscle

4
Growth
and
development

The processes of fertilisation and fetal development discussed in this chapter are illustrated in the television programme *Life Before Birth*.

Living creatures are faced in each new generation with a serious problem: how to produce new individuals with the same degree of complexity as themselves. Single-celled creatures solve this dilemma by increasing in size and then dividing in two. There is no known mechanism by which multicelled animals can do the same thing. Some simple multicellular animals, such as *Hydra*, can 'regenerate' totally, even from small parts of the adult animal. Earthworms chopped in two regenerate a new 'tail end'. But regeneration of this kind has its limitations.

Even if it were possible for a human to regenerate in the same way, the impairment in the interim would be tremendous and would last over a long period. Regeneration does occur in complex animals, but only in a limited manner. A newt can 'lose' its tail when escaping from a predator, and the tail will regenerate, and a limited amount of regeneration will occur if the tip of a child's finger or toe is amputated.

Generally, however, animals do not use the phenomenon of regeneration after 'splitting' as their main mode of reproduction. Instead, they replicate by *sexual reproduction*, by the fusion of genetic material from two parents. This process seems at first to be the opposite of cell division.

□ How so?
■ Cell division involves the formation of two cells from one 'parent' cell. Sexual reproduction involves the formation of one cell as a result of the fusion of two cells, the male sperm and the female ovum.

Why sex?
When describing mitosis in Chapter 2 you saw that replication of DNA within the cell is necessary before cell division occurs, so that each daughter cell gets a copy of all the DNA present in the parent cell. However, ova and sperm cells are made by a process called *meiosis* in which they get only *half* the DNA from the parent cell. Then, when DNA from ova and sperm combine at fertilisation, the normal amount of DNA is restored. Why go through

this complicated process to get back to the starting point? The value of sexual reproduction is that it increases the potential range of variation within a population.

☐ Why might variation be an advantage to a species?

■ As you saw in Chapter 1, a species evolves by the process of natural selection favouring those individuals who are best *adapted* to the environment. If the environment never altered, then there would be no advantage to a species in producing variations in its offspring. However, environments do change with time. Thus, variation increases an individual's chance of leaving offspring that will have the ability to survive in the changed environment.

Variation arises not only from the mixing of genes from each parent during meiosis, but also from random changes occurring in single genes or groups of genes in the DNA of the sperm and ova. These changes generally take the form of alterations in the sequence of bases in the DNA chain that forms a gene. They are known as *mutations*. Their cause is largely unknown, although it is known that some chemicals and some forms of radiation, such as X-rays, can affect the DNA in this way.

☐ Would you expect the mutation of a single base to have any effect on the protein ultimately made from the DNA template containing the changed gene?

■ Probably yes. As you saw in Chapter 2, proteins are chains of amino acids. A change in even a single base in a gene can lead to a change in the amino acids incorporated into the protein being synthesised.

Sometimes mutations are lethal to the creature. However, in others the changes may be so small that complicated biochemical tests are needed to identify them.

As you have seen, each chromosome is one of a pair. Apart from the genes on the sex chromosomes, each gene has an opposite partner on its paired chromosome. Thus an individual possesses two 'instructions' for each protein it synthesises. For example, there are two genes that instruct what colour the eyes will be. These different versions of the same gene are known as *alleles*. In this example, depending on which alleles of the gene are present, a person may have brown or blue eyes. Sperm and ova cells each contain one allele of the gene coding for eye pigment. When they fuse, the offspring cell has an allele from each parent. If the alleles are the same, the individual is known as homozygous. Thus if both alleles code for brown pigment the offspring's eyes will be brown. If, however, the offspring contains a 'brown' and a 'blue' allele, a condition known as heterozygous, what colour will the eyes be? Are there rules for predicting the eye colour of an offspring or do the colours just mix? It turns out that there *are* such rules and we shall discuss these shortly.

So, genetic variation in humans is produced by the mixing of alleles from sperm and ova, and by mutation. Sperm and ova cells each have only one set of the chromosomes present in the rest of the parent's body cells, and the DNA in the chromosomes is 'shuffled' so that ova and sperm will have differing combinations of the alleles present in the parents' body cells. Thus each ovum and each sperm has a unique assortment of the parental alleles, and offspring end up with a unique set of genes, known as their genotype. This means that a great variety of genetic material is expressed, and interacts with the environment. This increases the chances of an individual leaving offspring as it increases the chances of successfully adapting to changing environmental conditions.

There is just one problem left unresolved by this neat scheme. It may explain *why* sex, but it does not explain *how* sex. We have said that males and females have the same number of chromosomes, twenty-three pairs. Of these, twenty-two pairs (autosomes) are similar in males and females. The twenty-third pair (sex chromosomes), however, is not: in females there are two so-called X-chromosomes, whereas in males only one of this pair is an X-chromosome, the other is a Y-chromosome.

☐ What sex chromosome will the male's sperm contain?

■ Half the sperm will carry an X-chromosome, the other half a Y-chromosome.

☐ So what are the possible results of fusion between an ovum and a sperm in terms of the sex chromosomes?

■ If an 'X sperm' meets an ovum (which must be X), the fused cell will be XX (a daughter). If a 'Y sperm' meets an ovum, the fused cell will be XY (a son).

So the existence of the sex chromosomes ensures that sexual differences themselves are reproduced in the next generation. The sex chromosomes carry other genes in addition to those for sex determination. As you will see, this has important implications for certain types of disease and genetic malfunction.

Fertilisation

Conception occurs when an unfertilised ovum and a sperm unite. When the head of the sperm, which consists mainly of the sperm nucleus, enters the ovum and fuses with the nucleus of the ovum, its tail falls off. Ovum and sperm each have one copy of every chromosome, so the nucleus of the fertilised ovum contains a full set of chromosomes (twenty-three pairs in humans) that can be considered as a combined set of instructions that will help create a new multicellular organism from a single cell (Figure 4.1).

Also contained in the fertilised ovum, in the cytoplasm of the cell, is a second set of instructions in the form of

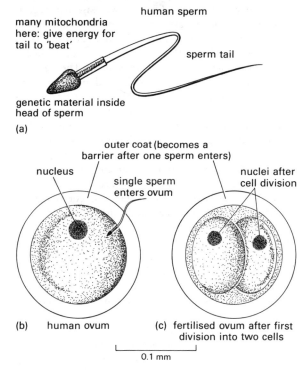

many mitochondria here: give energy for tail to 'beat'

human sperm

sperm tail

genetic material inside head of sperm

(a)

outer coat (becomes a barrier after one sperm enters)

nucleus

single sperm enters ovum

nuclei after cell division

(b) human ovum

(c) fertilised ovum after first division into two cells

0.1 mm

Figure 4.1 (a) Human sperm. (b) Fertilisation occurs when there is fusion between an ovum and a single sperm. (c) The first cell division.

chemicals that allow the first few hours of cell division and multiplication to take place without the need to 'ask the DNA in the nucleus for instructions'. These chemicals are packaged into the ovum during its formation in the ovary. The exact number and type are not known, but the most important seem to be proteins and some types of RNA.

☐ Can you imagine why this pre-packaging might be necessary?

■ The earliest stages of development involve a 'hurly burly' of frequent division and cell specialisations. If the dividing cells had to wait for DNA to produce RNA, and RNA to produce proteins, development would be slowed.

We have looked at fertilisation as a rather automatic process, but you should not underestimate the journey that the two participating cells have to make in humans. A discussion of the male and female sex organs will be found in Chapter 10 but general knowledge gives you enough background for present purposes. The sperm has to travel from the penis up through the womb and into the Fallopian tubes, a distance that may be as much as 20 centimetres. The ovum is about 0.1 millimetre in diameter. The diameter of the sperm's head is one-tenth of this, that is, about 0.01 millimetre. There are, therefore, immense

problems of navigation and timing in getting the two participants together, involving the equivalent of a person walking as much as 40 km. The sperm is aided in its search by chemicals sent out from the ovum. This maximises the chances of a sperm reaching the ovum. In addition, millions of sperm are ejaculated together, again increasing the chances of at least one making the long journey successfully.

The genetics of heredity

Once the ovum and sperm have fused, the construction of the new individual can begin. Just what this new individual will be like, of course depends on just which alleles of the parental genes are present in the particular chromosomal combination the individual has inherited. Geneticists try to relate particular aspects of the individual — height, weight, eye colour and so forth, sometimes referred to as particular *characters* — to the presence or absence of particular combinations of genes. Some characters, like height, are affected by many genes within the chromosome, so that in general the presence or absence of a single gene has no major effect. However, other characters are very sharply and straightforwardly affected by the presence or absence of single genes. For example, many genes code for enzymes. If the 'normal' allele of the gene is present, a normal amount of the enzyme will be made, but if an 'abnormal' allele is present the enzyme may be missing altogether, or produced in very low quantities. The result *may* then be a single specific change in the 'character', such as in the example we gave earlier of eye colour.

Eye colour is an example of a genetically determined character which varies within the human population, but without obviously harmful or beneficial consequences. Other genetic differences, however, may, as you will see, be associated with specific diseases.

☐ Why would you expect there to be single genes coding for a protein, but not single genes coding for height or weight?

■ Because there may be thousands of different body processes, involving enzymic reactions, which all help determine weight or height. Presence or absence of a single enzyme might have little effect. An enzyme is a single protein and so may be coded for by one gene (though some enzymes are present in the body in multiple protein forms, each of which is coded for by a different gene).

As the offspring has one copy of each of its genes from each parent, and each gene may exist as one of two alleles, what determines how these alleles are 'expressed' in the developing embryo? Let us take, as a hypothetical example, a gene G that produces the blood protein haemoglobin. This gene can exist in a different form—let us call it g—

and in this form it produces an abnormal protein known as sickle haemoglobin.

☐ In a population of people, some of whom have the sickle haemoglobin gene (g) and some do not, what combinations of genes might be present in the two parents of a child?

■ Each parent has two copies of the gene, and the possible combinations are GG, gg, or Gg.

☐ What alleles could be present in the sperm of a man who has GG?

■ Only G.

☐ Suppose the man has one G and one g copy of the gene, what alleles will be in his sperm?

■ Each sperm will be either G or g. Similarly, the ovum from a woman who has G and g will be either G or g.

The offspring of a GG woman and a GG man will obviously be GG. Similarly, the offspring of a gg woman and a gg man will be gg. A GG man and a gg woman can produce only Gg offspring.

☐ If both parents are Gg, what are the possible combinations in the offspring?

■ Clearly, GG, gg and Gg are all possible.

☐ Is there any way of predicting which combination will occur in any particular offspring?

■ No, the alleles are randomly distributed during the production of sperm and ova, and which sperm finally fuses with which ovum is also random.

However, it is possible to say in what *proportion* the different combinations will occur.

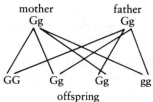

mother   father
Gg   Gg

GG   Gg   Gg   gg
offspring

It makes no difference to the offspring whether it gets its G or g from the mother or the father, and therefore the genes for haemoglobin of the two Gg offspring are identical. This combination will therefore be twice as frequent as either GG or gg. Thus 1/4 of the offspring will, on average, be GG, 1/4 will be gg, and 1/2 will be Gg.

What does this say about the type of haemoglobin the offspring possesses? Obviously, the GG child will have normal haemoglobin, and the gg child sickle haemoglobin. What about the Gg combinations? It turns out that when a person has a pair of different alleles of the same gene like this, one of them is normally expressed more than the other, sometimes completely so. The allele which is expressed is described as *dominant*, the other as *recessive* (and

conventionally represented by a lower-case letter, g, as opposed to an upper-case letter, G, for the dominant). In the case of sickle haemoglobin, the normal form (G) is dominant and all the offspring which have at least one G will have virtually normal haemoglobin. The gg form makes sickle haemoglobin and a child born with this combination will suffer from sickle-cell anaemia. Under most, though not all, conditions both Gg offspring will be indistinguishable from normal offspring, but they will be *carriers*, like the parents. They will have a 1 in 4 chance of themselves producing offspring with sickle haemoglobin.

The rules of transmission of characters by single genes and the relationship of dominant and recessive characters were first worked out by a Czech monk, Gregor Mendel, in the 1860s, but the molecular mechanisms involved in the transmission were not understood until 100 years later, in the 1950s and 1960s, when the mechanisms of the DNA code were worked out. 'Mendel's laws' are often seen as the foundation stone of genetics. Nonetheless, it is important to understand that the inheritance of most 'characters' is much more complex than these simple rules suggest. Thousands of genes may be involved in a character such as height and, although there may be a clear understanding that genes play a part in the process, just *which* genes, and the hereditary relationships involved, is a much less easy matter to understand. In addition, the genetic transmission of characters is not merely a matter of the presence or absence of particular genes, whether a single gene or many. The effect a gene has on the development and health of an individual will depend on the environment in which that person exists. An example of the importance of the environment will be discussed in Chapter 12.

Developmental processes

As the fertilised ovum divides, a mass of cells, the developing embryo, is produced. (We shall use 'embryo' to refer to all stages before birth, and the term fetus to refer to a human embryo from the time that it is recognisably human in appearance.) Development consists of a number of processes that go on at the same time. The most important of these processes are cell division, cell differentiation and morphogenesis. We shall discuss each of these in turn.

Cell division is the process by which the single-celled ovum generates the six million, million cells of the newly born human baby. Although the first few cell divisions are synchronised (that is, the ovum divides to give two cells; both these cells divide to give four cells and so on), later divisions occur in a less regular manner. Cells in one part of the developing body may divide twice as often as those in another part.

Cell differentiation refers to the acquisition of specialisations by cells as they divide. The *structure* of different cell types is specialised in a way suited to its particular *function*. Nerve cells not only look very different from muscle cells, but they perform very different tasks within the division of labour shown by multicelled creatures.

   ☐ How would you expect a cell in an embryo of only, say, 64 cells to differ from a cell in an embryo of the same type of animal at a later stage of development, say when 50 000 cells are present?

   ■ A cell in the 'younger' embryo would be much less specialised. General characteristics emerge first during development, so whereas a cell in a 50 000-cell embryo might belong to one of dozens of specific types ('liver' or 'kidney', say), a cell in a 64-cell embryo might be one of only three or four basic 'cell types'.

Morphogenesis is the process whereby an embryo or part of an embryo takes on specific shapes and patterns as a result of the interaction of the processes of cell division and cell differentiation. The rules which govern morphogenesis are not yet known and there is still little firm knowledge of how patterns are set up in growing tissues. It is probable that there are several different factors, and it appears that one factor might be the chemical differences that exist between cells that are in different positions.

Developmental events

The first few divisions of the ovum involve some simple steps. Each division (called cleavage) roughly divides the existing cells into two, so 2, 4, 8, 16 and so on cells result from sequential divisions (Figure 4.2). The DNA in the nucleus has to divide and replicate itself before each cell division, and the cells grow by absorbing raw food material from the surrounding environment and from the 'lunch packs' of protein and RNA provided in the cytoplasm of the ovum itself. The earliest food is obtained by diffusion of fluids from the wall of the womb into the embryo.

   As cleavage proceeds, the cells become arranged around a hollow cavity, the cell layer being thickened on one side. This is the first stage in a progressive series of specialisations that lead not only to the growth of the embryo, but also to the formation of the intimate links between mother and embryo necessary to maintain the embryo's 'life-support systems'. These life-support systems are fairly complex. We shall consider them further in Chapter 10. For the present, let us consider the embryo itself.

   Development involves progress from single cell to adult animal. The procedures used are rather like those artists might employ when painting a picture. They first sketch in the rough outlines of the major features and only then work up to finer and finer detail. All animals use the same basic schedules to produce their own work of art. In animals such as frogs and fish it is easy to see the events of morphogenesis because growth occurs within a transparent egg membrane.

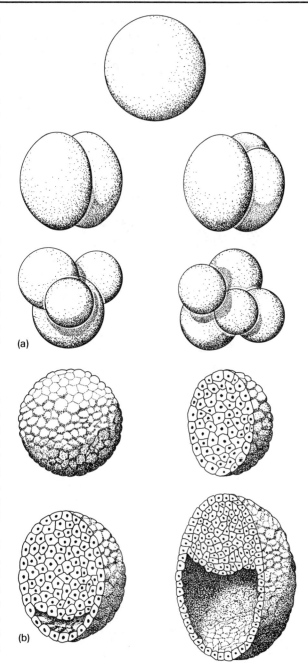

Figure 4.2 Embryogenesis: (a) early cell division (cleavage) and (b) the development of a cavity and a thickened cell plate which will form the embryo.

In humans, these events are both less visible and more complex.

   The first body system to form is the gut. This is produced by the pulling on and pushing in of one of the sides of the hollow ball of cells, in the same way as a hollow rubber ball can be deformed by pressure. The place where the push occurs eventually becomes the anus and the end of the internal projection breaks through the other side of the embryo to form the mouth. The digestive tube is then mapped out. After the gut starts to push itself inwards, the earliest signs of brain and spinal cord appear. Cells along the top of the embryo 'tuck in' to produce a neural tube (Figure 4.3). The cells which make up this tube then produce, at the front end, the brain and those further back make up the spinal cord. At the same time as the brain is forming, other types of cells and tissues form.

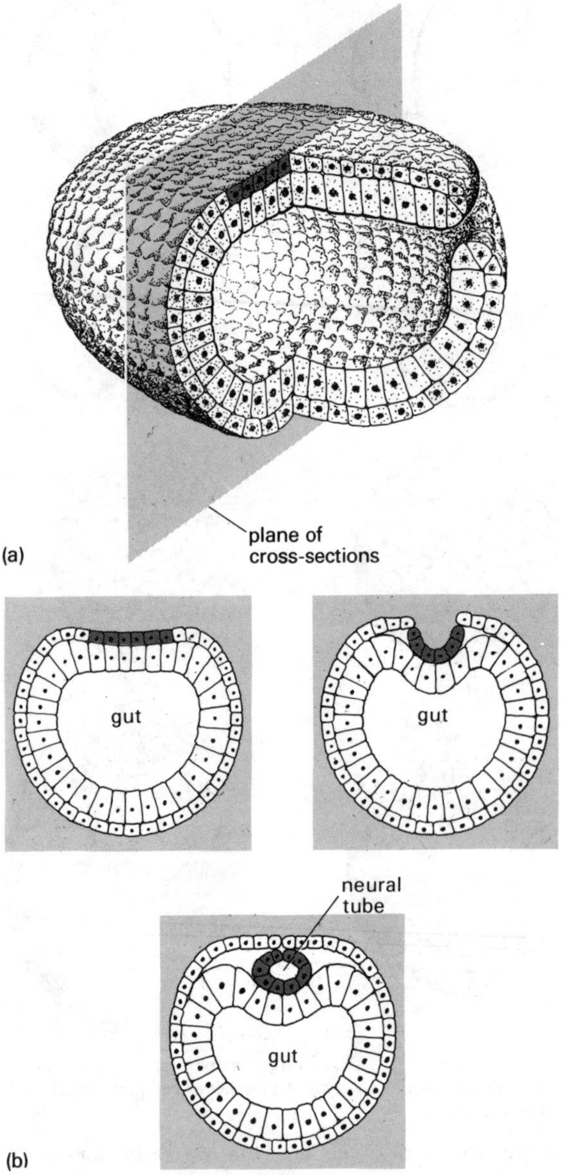

As development proceeds, certain general properties that are possessed by all cells of the early embryo are lost. Early in development, cells can still move around quite freely, migrating past others to reach 'final' positions according to rules that are still quite unknown to us. Later this property is lost, as is the capacity for frequent cell division. Adult populations of cells divide slowly, and some (such as the nerve cells of the brain) not at all. In addition, and perhaps most importantly, cells become specialised. As you know, all the cells of the body contain all the DNA necessary to code for every protein the body makes. Yet individual adult cells differ not only in characteristic shape but also in their protein composition. In each a rather different set of the genes are 'switched on'. Development thus must also involve a process whereby generalist cells become specialist. Immediately after fertilisation, virtually any cell in the ball of dividing cells can become a nerve cell, a muscle cell or whatever. Within a small number of divisions, though, the fate of the cell is sealed and it is committed to becoming a particular type. This process of specialisation is important both in itself and because there are circumstances (as you will see in relation to cancer) when specialised cells appear to reverse this process and become generalist and rapidly dividing — 'de-differentiated'—once more.

Growth after birth

The human individual has the longest period of development after birth of any animal. Physical growth continues until the late teens, with accelerated increases in the fetal and adolescent periods, and slower growth during infancy, the juvenile years and after adolescence (Figure 4.4). Little extra tissue is added after the age of twenty, and any subsequent change in weight is due chiefly to the addition or reduction of fat, changes in water level in tissues and changes in the amount of bone present. Throughout life, however, cells are continually dying and are being replaced by new ones. At birth, the majority of the body

(a)

plane of
cross-sections

(b)

neural
tube

gut

gut

gut

Figure 4.3 Formation of the neural tube in the developing embryo: (a) at four weeks the embryo is a hollow structure, two cells thick; (b) consecutive cross-sections during the fourth week showing the development of the neural tube.

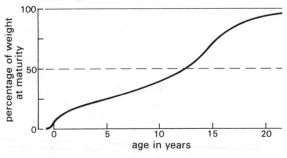

Figure 4.4 Growth from conception to adulthood.

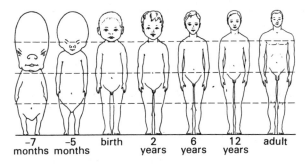

-7 months   -5 months   birth   2 years   6 years   12 years   adult

Figure 4.5 Alterations in body proportions from fetus to adult.

structures are formed, with the exception of the replacement of some cartilage by bone and the development of sex organs at puberty.

☐ Look at Figure 4.5. Make a note of some of the progressive changes which you can see occurring with age.
■ Proportions of most body parts seem to change with age. The most noticeable difference is the relative size of the head, legs and arms.

The changes in body proportion are brought about by local changes in cell division as a result of various cues. Some of these changes are undoubtedly due to the kinds of cell interactions that act during fetal development. Certain chemicals also play a major role in regulating growth.

Cell ageing

We are all aware that a baby's skin differs from that of a 6-year old or a 60-year old. Similar changes take place in all tissues of the body throughout the life cycle. What alterations are taking place in the cells to account for these changes? There is no general agreement about the fundamental cause or causes of ageing. It does seem that several or even many different factors are involved. Cell death (unless made good by new cells), or impairment of cell function, will have far-reaching effects right up the biological hierarchy to the whole body. Ageing effects are

seen in the chemicals of the body such as lipids and proteins: as the body ages, the proportions of different lipids and proteins within the cells change. This accounts for the different taste of lamb or veal from mutton or beef, for instance. Bone structure changes as bones become more fragile; tissues like skin lose their elasticity; and other cell properties change.

The clue to ageing seems to lie, paradoxically enough, in cell division. *Too many* cell divisions, to be more accurate. It would be a difficult task to work out how many cell divisions have occurred since a 70-year-old liver began its life as a few cells in an early embryo. Each time a division occurs, the chromosomes have to be copied and passed on to the daughter cells. At each copying step DNA bases may be damaged or miscopied. Most of the miscopies are eliminated by special enzymes, but accumulated errors still occur. Other chemical degenerations may also occur. Defects in the enzymes controlling DNA replication can lead to faulty DNA and, hence, faulty chromosome copies and defective proteins may be produced. As a result, the working of the cell's components may become inefficient and the cell's function may be seriously impaired or it may even die.

Cells are also affected by their environment, which can act at a series of levels. For instance, a particular diet may lead to more rapid decay of teeth which might itself lead to malnutrition. The components each cell needs for its efficient working might then be absent. Radiation in the environment can increase the amount of damage to DNA and other chemicals in cells.

A further factor in cell ageing is the influence of the genetic material, the DNA itself. It is thought that actual DNA base sequences may affect both the probability that particular types of damage occur and also the extent of any repair, compensation or replacement that may take place. Limits to the number of divisions that cells can undergo may be of benefit to animals. If a cell gets out of its normal place in the body, it might escape the sequence of controls on its behaviour and could divide in a disorganised fashion. It should be added, however, that this idea has little supporting evidence at present.

Objectives for Chapter 4

When you have studied this chapter, you should be able to:

4.1 Explain the evolutionary advantage of sexual reproduction and how genetic variation between individuals arises.

4.2 Describe the way in which alleles are assorted during sexual reproduction.

4.3 Discuss the four key features of development.

4.4 Explain the cellular changes that take place with ageing.

Questions for Chapter 4

1 (*Objective 4.1*) What feature of the structure of DNA limits the number of possible variations of a particular gene that will code for a useful protein, and why?

2 (*Objective 4.2*) An albino (someone lacking pigment in their skin) has alleles *aa* for the albino gene. His 'normal' wife has alleles *AA*. If the allele *a* is recessive, would you expect any of the children to be albino and, if so, in what proportion? If the man had been married to a woman with alleles *Aa*, what proportion of the children would be albino?

3 (*Objective 4.3*) In what main ways does a nerve cell in the adult (a) resemble, (b) differ from, any of the cells in the sixteen-cell embryo?

4 (*Objective 4.4*) To what extent and in what way is ageing dependent on cell division?

5

On being an organism

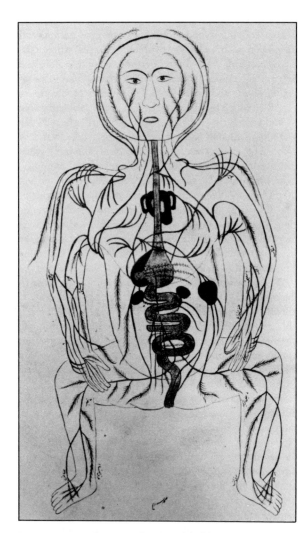

Persian seventeenth century drawing of the blood system and gut.

So far we have concentrated for the most part on the basic 'units' from which all living things are made—cells. You have learnt something about their structure, the biochemical reactions that go on within them and the mechanisms by which DNA initiates and regulates all cellular processes. You have seen that cells differentiate and that, in multicelled creatures, groups of specialised cells form tissues and organs. In this, and the following five chapters, we shall move up the hierarchy of biological organisation and focus on the physiology of the human body, paying less attention to its constituent cells than to the interacting systems of tissues and organs on which our lives depend.

☐ Can you recall what are the six characteristic processes that distinguish living things from non-living material?

■ All organisms feed and expel the waste products of digestion; move independently of external forces; respond to changes in their environment; grow; repair themselves when damaged (at least to some extent); and reproduce so that a new generation replaces the previous one.

The way in which different types of organisms achieve these activities varies. Humans, in common with all other mammals, have certain distinguishing physiological features. Mammals all strive to maintain a fairly constant body temperature, no matter what the temperature of the outside world, and give birth to young which have developed inside the female's body, and which she then suckles with milk produced by special glands.

There are significant advantages in having a relatively large body mass, being able to move freely from one place to another, and maintaining a constant body temperature. We discussed benefits of these three features in Chapter 3.

☐ What are these benefits?

■ An organism composed of many millions of cells can use groups of them to fulfil specialised tasks that are far more complex and sophisticated than those within the scope of a single-celled or very small multicellular creature.

Freedom of movement enables food, shelter and sexual partners to be sought after, and the avoidance of predators and extremes of environmental conditions.

Maintaining a constant temperature opens up a wider range of habitats and the ability to keep going during daily or seasonal changes in the temperature of the outside world.

This last point needs further elaboration. All chemical reactions (be they biochemical ones inside cells, in the formation of rocks, or in the laboratory) work best at a particular, or optimum, temperature. This means that more of the product of the reaction is formed in a shorter time than at temperatures above, or below, this optimum. Animals such as reptiles that cannot control their body temperature, but let it drift with the environment, cannot perform biochemical reactions within their cells as efficiently as animals that can keep their temperature constantly close to the optimum. In mammals, biochemical reactions and control of body temperature have both evolved and adapted to one another over millions of years. In humans the optimum temperature at which most reactions take place is 37 °C (or 98 °F).

However, even though the evolution of a constant body temperature has advantages, it also poses a serious problem. How can such a stable state be maintained during summer and winter, during strenuous exercise and at rest? Complex *control mechanisms* have had to evolve, which regulate body heat and distribute it fairly evenly around the body. We shall discuss this process in detail in Chapter 8.

   □   Can you think of any other problems posed by a large body mass?
   ■   Cells need a constant supply of nutrients, particularly glucose, amino acids and certain salts, and oxygen, in order to stay alive. They must also get rid of poisonous waste products such as carbon dioxide. In very small creatures molecules can simply diffuse in and out, or be pumped across the cell membrane to or from an external environment which surrounds the cell. An animal with many millions of cells cannot interface all of them with the outside world, and so must evolve systems for delivering 'essential services' to all parts of the body and for expelling waste products. Moreover, the supply of nutrients must be relatively constant.

This requires an extensive transport network carrying nutrients and oxygen to all cells and removing waste. Even rather simple animals such as earthworms have solved this problem in much the same way as humans, by evolving a network of fine branching tubes, filled with fluids, special 'carrier' molecules and cells, that penetrates every part of

the body. A pumping mechanism keeps fresh supplies circulating and delivers waste to detoxification centres (in humans, the liver and kidneys). The circulatory system of humans is the subject of Chapter 7, and in Chapter 8 we shall discuss how the supply of nutrients and oxygen is regulated and how waste products are excreted.

The circulatory system has another important function in maintaining health. It is inevitable that injuries will be sustained during life—burns, cuts, bruises and so forth—and these must be repaired. Before this can happen the breach in the body must be sealed to prevent loss of blood and damaged tissues must be removed so that new growth can occur. In addition, the injury may allow the entry of microorganisms that could disrupt normal functions, perhaps so severely that death might result. The circulatory system contains blood clotting mechanisms and several different types of cell that are specially adapted to digest the debris resulting from injury, promote wound healing and destroy infectious organisms. The linked processes of *inflammation* and *immunity* are the subject of Chapter 9.

All of which brings us to the nub of the problem posed by multicellularity. If all these basic needs are to be met, day and night, summer and winter, asleep or awake, how can a constant, stable state be achieved *inside* the body when the outside world fluctuates in temperature, humidity, light, pressure and many other aspects? Add to this the fact that a human's *interaction* with the external environment also fluctuates (large but occasional meals, sleeping for eight hours out of every twenty-four, performing strenuous exercise sporadically, etc.), and you can see that very sensitive regulation of the conditions inside the body must have evolved in order to keep them relatively constant.

The regulation of these internal conditions at close to the optimum level for each organism, as you will recall, is known as homeostasis. Note that homeostasis is a *process*, an active balancing of physiological systems such as the circulatory or excretory systems, in which the internal conditions are constantly monitored and adjustments are made to keep vital parameters like body temperature close to the optimum level. The concept of homeostasis is central to modern physiology. Over the past 100 years or so the human body has come to be seen as a complex network of interacting systems, constantly monitoring and adjusting the internal state. This 'systems' view incorporates numerous metaphors from the world of systems technology. Thus the regulation of body temperature is seen as analogous to the way in which an automated central heating system works: a sensing device is necessary to sample temperature and compare it with a pre-set optimum level and, if the sampled temperature deviates from the set-point, a *feedback* loop is activated and appropriate adjustments made. You will meet such metaphors many times in what follows. However, no matter how useful these ways of describing physiology have proved to be, you should remain aware that fashions in biological metaphors have changed many times in the past and may do so again.

It is also worth noting that the concept of homeostasis does not only apply to the 'whole body' level of multicellular animals. Each individual cell is also maintaining homeostasis, in the sense that nutrients and oxygen must be kept at a level high enough to sustain life and waste products cannot be allowed to rise beyond certain critical concentrations. The action of enzymes regulating the biochemical reactions inside cells is the major mechanism available to maintain a cell's internal environment close to optimum conditions. More complex, multicelled organisms — particularly animals — have evolved very sophisticated control systems to maintain homeostasis even in hostile external environments.

These control systems are the subject of Chapter 6 and consist of the nervous system (the brain, spinal cord and nerves) and the hormonal system (the many glands in the body which secrete *hormones*, chemicals that are transported in the blood and which alter the function of cells, tissues and organs in precise ways). Together these two systems perform the highly sophisticated task of regulating the internal state of the body. Of course, the nervous system does much more than this. It enables us to perform complex tasks of reasoning and logic, to learn and memorise information, to collect data about the outside world, as well as the state of our body, and to coordinate complex actions. In humans at least, 'higher' functions such as consciousness seem to reside in the brain, together with the ability to produce spoken and written language, compose music, paint pictures and much more.

Some hormones also have functions beyond the control of homeostasis, particularly in relation to the production of new offspring. In Chapter 10 we turn to the process of sexual reproduction in mammals and the crucial involvement of the so-called sex hormones.

You may have noticed that we have been flipping backwards and forwards during the course of this discussion, introducing topics that will be explained fully in one or other of the five chapters to follow, one moment talking about Chapter 8, then 7, on to 10, back to 6, and so on. This is not simply chaotic thinking on our part! In human physiology everything links up to everything else. Thus, explaining the regulation of heart rate involves the structure of the heart itself, sensors that monitor the amount of oxygen and carbon dioxide in the blood, electrical impulses from the nervous system, and the hormone adrenalin. Where should one begin?

There is no perfect answer, so we have chosen the following strategy. Physiology is mainly concerned with homeostasis, the regulation of the internal state. Therefore you cannot understand much about the healthy functions of the body without knowing something about the nervous and endocrine systems. So we shall discuss them first, in the next chapter. Then we shall move on to describe the transport network provided by the circulatory system (Chapter 7). With these fundamentals of regulation and transport as a firm foundation, we can then look at the regulation of body temperature, oxygen levels, blood sugar, and toxic waste in Chapter 8. Chapter 9 is about inflammation and immunity, and reproduction is the theme of Chapter 10.

Finally, as you study these chapters, notice that the systems and processes we describe are subject to significant variations. The heart and circulation of a newborn baby are not identical with those of a 5-year-old child, an overweight adult, an athlete or a person in extreme old age. Even people of similar ages and lifestyles show significant variations in their physiology, just as their faces and fingerprints differ. The amount of air they take in with every breath, their pulse rate at rest, even their body temperature, all will show individual variations. There is a great temptation to discuss physiology as though human beings are identical but, like most things in life, it is not that simple!

Objective for Chapter 5

When you have studied this chapter, you should be able to:

5.1 Discuss the problems posed by multicellularity and, in simple terms, outline the solutions illustrated by the human body.

Question for Chapter 5

1 (*Objective 5.1*) Imagine a muscle cell embedded several centimetres below the skin, in the centre of a large muscle in, say, the thigh of an athlete. Describe, in general terms, how the basic requirements that sustain life are met for this cell and how this differs from the processes available to a bacterium.

6
Control and communication

Illustration by Alexander Monro (1783) of nerves entering striped muscle.

In Chapter 5 much was said about homeostasis and the control of body processes by complex systems involving sensing and sampling devices, feedback loops and 'actions' taken within the body to maintain key parameters close to optimum levels. Control mechanisms in the body exist at a number of levels. At the earliest stage of development the myriad cell divisions in a young embryo may be controlled by the inbuilt genetic 'programme' of the cells interacting with chemical signals received from neighbouring cells. On the other hand, control of a complex function such as the rate of heartbeat involves the interaction of electrical signals from the nervous system, the concentration of certain hormones in the blood, and the inbuilt capacity of the heart muscles to respond to these signals.

In humans, the control of heartbeat, the processes of breathing and digestion all involve such coordination, but they are all below the level of what is described as consciousness. Other aspects of human activity are controlled and directed consciously in voluntary movements such as walking and running, perception of sensory information, speaking and writing. Yet the mechanisms involved in such conscious activity must also be amenable to biological explanation.

We cannot hope to unravel the complexities of all these processes in the limited space available, but we can make a start by describing the basic features of the two main control systems in the body—the nervous system and the hormonal system. These two systems collaborate in regulating many aspects of the internal state: for example, temperature, blood pressure, oxygen and glucose levels. All the glands which synthesise hormones have nerves running to them. The release of hormones from the glands is (at least in part) subject to control from the nervous system. Conversely, all parts of the nervous system are in close contact with blood vessels. Hormones are transported in the bloodstream, providing a means by which the level of hormones in the circulation can affect the activity of the nervous system.

Both the nervous system and the hormonal system modify the activity of cells, tissues and organs throughout the body. Why have two such systems? One answer is that the type of modification that each can induce is qualitatively and quantitatively different. The nervous system works on electrical principles, is quick acting and used for sensing and providing control of events which require rapid response. Nerves terminate in precise locations, so their activity can provide pin-point control, for example, of movements. The hormonal system, relying as it does on production and release of chemical substances, is slower acting, but can induce a sustained effect more 'economically' since these chemicals are diffused throughout the whole body.

There is good evidence that the hormonal system evolved in multicellular organisms relatively early. To coordinate the activity of many cells, there must be some way that the cells at one end of an organism 'tell' those at the other end what is going on. This can be achieved by the secretion of a chemical 'messenger' which can diffuse through the spaces between cells, carrying the information as it goes. Plants as well as animals make use of such hormonal signalling systems to, for example, initiate growth.

Nervous systems, which are ways of getting information very quickly and precisely from one set of cells to another, bypassing everything in between, are, of course, confined to animals which need to be able to respond and act quickly on their environment. They came later in evolutionary terms. The first true nervous systems appeared in worms and the first brains only after creatures with backbones (vertebrates) had evolved.

As you will see, nervous systems make use of chemical mechanisms which were 'pioneered', in evolutionary terms, by hormones. In deference to their evolutionary precedence, we begin our discussion with hormones before turning to the nervous system.

Hormones

Hormones are chemical substances that are synthesised and secreted by specialised cells. They are transported in the blood to other parts of the body where they exert control on the physiological activity of the 'target' tissues. Some hormones exert only very local control on highly specific cells. For example, the presence of food in various portions of the gut triggers the release of a hormone that stimulates the next portion of the gut to contract or secrete digestive enzymes. We shall not discuss these local hormones further here, but shall concentrate on those hormones with major, generalised effects on the body.

How do hormones exert control over the functioning of target tissues? First, hormones affect the basic metabolism of the body, that is, the rate of chemical reactions, transport of substances across cell membranes, the rate of cell divisions during growth and development and the secretion

of a huge number of important substances such as enzymes in the digestive system.

Second, a few of these hormones affect virtually *all* the cells in the body, tuning the metabolism of every cell. Others affect only specific target tissues, but these are of major significance: for example, regulating the ability to produce ova or sperm, sustaining a pregnancy, or controlling the filtration of toxic molecules from the blood.

Third, a few hormones are produced at a more or less constant rate, at least during certain parts of the life cycle. Others show a rhythmic secretion pattern, fluctuating in response to certain cues. Insulin levels, for example, fluctuate in response to glucose in the bloodstream and thus rise and fall several times a day. Other hormones show a diurnal (day–night) rhythm, and others, such as those controlling the menstrual cycle, have even longer cycles.

Where are hormones produced in the body? Hormones are secreted by endocrine glands, groups of specialised cells that synthesise a particular hormone. The hormone is released from the gland directly into the bloodstream. Positions of these glands are shown in Figure 6.1. The pituitary, adrenals and thyroid are distinct and separate structures, whose sole function is to produce hormones. On the other hand, hormone-producing cells embedded in the

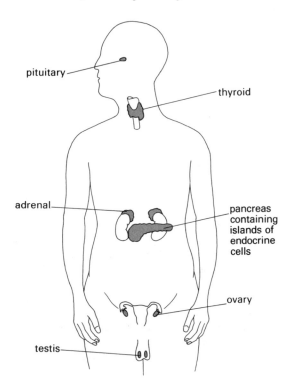

Figure 6.1 The location of the main endocrine glands. (Artistic licence has been used to show both the position of ovaries and testes!)

pancreas and the testes or ovaries are only part of the organ, which has other functions to perform. In addition, the placenta acts as a source of hormones during pregnancy. All the endocrine glands tend to overproduce, so their effects generally 'overshoot' slightly. This small excess effect is enough to activate signals which feed back to an endocrine gland and cause its output to reduce or cease altogether. Such *negative feedback* (feedback that leads to a reduction rather than increase in output) is a common feature in the regulation of the endocrine glands.

The pituitary

Several hormones are of major importance because they regulate the output of other hormone-producing tissues. The main endocrine gland responsible for regulating other glands is the pituitary, so much so that the hormonal system has often been likened to an orchestra with the pituitary as the 'conductor'. In its turn, the pituitary is almost totally under the control of a nearby brain region called the hypothalamus. Thus even the 'conductor' ultimately comes under the control of the nervous system.

The pituitary produces eight major hormones and several minor ones. Names of the major hormones are given below in brackets and are for reference purposes only, we don't expect you to memorise them. You should note, however, the very wide range of effects exerted by this tiny gland, weighing less than 1 gm. Pituitary hormones regulate the output of the thyroid (thyroid-stimulating hormone), the adrenal glands (adrenocorticotrophin) and the gonads (gonadotrophins, of which there are two).

In addition to producing hormones to stimulate the other endocrine glands, the pituitary also produces hormones that control general growth and development (growth hormone), the development of breast tissue and lactation (prolactin), the rate of urine production (antidiuretic hormone — ADH) and uterine contractions during childbirth (oxytocin).

Thus the pituitary hormones directly or indirectly control many aspects of cell metabolism, growth, development and reproduction. A common cause of malfunction of body metabolism is a tumour in or near the pituitary gland, causing either an increase in output (because hormone-producing cells are multiplying in the tumour) or a decrease (because the nearby tumour is squashing and obstructing the gland). This can have dramatic effects. For example, under- or over-production of growth hormone during childhood can lead to dwarfism or gigantism, in which a person of normal body proportions remains tiny in adulthood or grows abnormally tall.

The thyroid

The hormone thyroxine is produced by the thyroid gland in the neck, and increases the metabolic activity of almost all cells in the body. Thyroxine contains iodine, and about 50 mg of iodine are required in the diet per year for normal hormone production. Increase in the concentration of thyroxine in the blood leads to increased metabolic rate, a measure of the rate of utilisation of foods for energy. The mechanism by which thyroxine (and other thyroid hormones) exerts this profound effect on cells is largely unknown, but one possibility is that it increases the rate at which proteins are 'translated' from messenger RNA codes, and indeed increases the rate of production of mRNA itself.

☐ How could this increase in protein synthesis affect the general metabolism of the cell?
■ If any of the proteins produced are enzymes involved in biochemical reactions in the cell, then an increase in their concentration would 'boost' the occurrence of those reactions. (In fact, experiments have shown that at least 100 such enzymes increase in quantity in the cell in response to thyroxine.)

The adrenals

The adrenal glands, situated just above each kidney, also exert a widespread effect on the body. Each gland is composed of two distinct parts. The inner part secretes adrenalin, a substance sometimes referred to as the *'fight or flight' hormone*. Adrenalin prepares the body for fast, dramatic action, speeding heart rate, increasing the blood supply to the muscles and restricting the blood supply to the digestive system. Anyone who has had a 'near-miss' in a dangerous situation will be aware of the feeling of an adrenalin flush. As you will see shortly, adrenalin is a hormone whose action is involved in the functioning of the nervous system and, unlike the other adrenal hormones, the release of adrenalin is not under the control of the pituitary, but is regulated primarily by the nervous system.

The outer part of the adrenal gland (the adrenal cortex) produces at least thirty different, but chemically-related, hormones known collectively as corticosteroids. Some corticosteroids influence the transport of the ions of sodium and potassium across cell membranes, particularly in the kidneys. Others are involved with the uptake and utilisation of glucose, proteins and fats from stores within the body. Corticosteroid secretion rises in response to virtually any kind of stress: for example, shocks (emotional or electrical), intense heat or cold, infection, injury and in disease states. A common-sense explanation of this, which has by no means been proved, is that corticosteroid makes more glucose available for the energy-consuming reactions of fight, flight or repair. Paradoxically, however, high corticosteroid levels *reduce* the effectiveness of the immune system (as you will see in Chapter 9), thus jeopardising the body's response to infection.

All the corticosteroids show a marked diurnal rhythm, being at their peak level in the early morning and at their lowest in the late evening. This rhythm occurs automatically and may have something to do with waking in the morning and preparing for sleep at night, although cynics might suggest that levels rise in the morning in response to the shock of getting out of bed! Disruption of this cycle may partly explain the uncomfortable physical sensations induced by jet-lag.

□ Can you suggest a mechanism that would tell the adrenal glands what time of day it is? (You will have to think back quite carefully to the factors that regulate the output of these glands.)

■ Corticosteroids are released from the adrenal glands in response to another hormone produced by the pituitary gland. This in turn as we have said is subject to regulation by signals from a region of the brain called the hypothalamus. It is reasonable to presume that information about the time of day is processed in the brain and relayed via nerves, and then pituitary hormones, to the adrenal glands.

This discussion by no means exhausts all that could have been said about the role of hormones in the regulation of human physiological processes. However, it should provide an adequate basis from which to understand the extent and range of hormone activity, and also a basis from which to appreciate the severe consequences for health of malfunction of the endocrine glands.

Nerves and how they work

The nervous system is a specialised arrangement of cells which together have the property of receiving information from the outside world and from regions within the body itself, storing and processing that information, and then signalling instructions to the rest of the body as to how to act in response to the processed information. The system is, of course, built, like all other body systems, of a set of cells — nerve cells — with a highly specialised form and function. These cells themselves are grouped into two subsets, the central nervous system, mainly the brain and the spinal cord, and the peripheral nervous system, a range of nerves scattered across virtually the whole of the rest of the body. For both peripheral and central nervous systems, though, the key to how the system functions is the individual nerve cell, or neuron. And the key to how the neuron works lies in its electrical properties.

Figure 6.2 shows two typical neurons. Each consists of a cell body, which has all the usual structures you now know to expect. There are also, running off the cell body, a series of thread-like projections called dendrites, whose function, as you will see, is to carry information *towards* the cell body, and one much longer thread called an axon

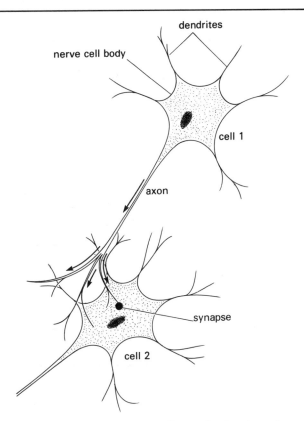

Figure 6.2 Neurons — nerve cells — showing how they interconnect.

which carries messages *away* from the cell body. The axon in the diagram is quite short, but in the human body some axons may run for many centimetres. The diagram also distorts the relationship between cell body, dendrites and axon, in that, in real life up to 80 per cent of the cell's volume can be in the dendrites and axon. The axon terminates either on another neuron (as in the diagram) or on a muscle or endocrine cell. At its termination, it breaks into a spray of fine processes, each of which makes contact with the membrane of the second cell in the form of a little swelling. The point of contact is called a synapse. Plate III(h) shows a stained microscope picture of nerve cells from the brain, magnified some 350 times.

How does the neuron work? Like all other cells, it is surrounded by a membrane and has a different internal *ionic* composition from that outside it. In particular, inside the neuron is a high concentration of the positively charged potassium ions (K^+) and charged ions of amino acids (in proteins) and other substances. Outside the membrane there are sodium ions (Na^+) and chloride ions (Cl^-).

The membrane is impermeable to large ions (such as those of proteins), but is semipermeable to small ions such as Na^+ and K^+. The membrane can change its structure so

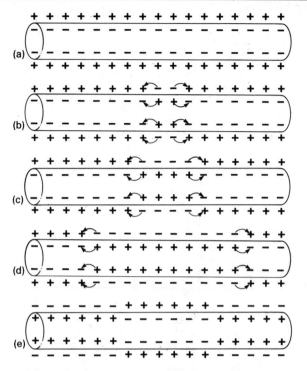

Figure 6.3 Transmission of an action potential (nerve impulse) along a neuron: (a) at resting potential; (b) the change in potential at the point of stimulus; (c) and (d) the impulse spreading away from the point of stimulus; (e) the resting potential restored as the impulse spreads away.

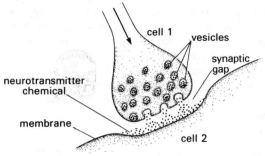

Figure 6.4 A synapse between two neurons.

in each direction, rather like ripples in a pool (Figure 6.3c and d). The impulse is no slouch, travelling at up to 120 metres per second—quite fast enough for the most adept space-invaders addict or squash player. Later, ATP is used up as the cell's sodium pump restores the ionic balance. Thus the resting membrane potential is restored, as at the centre of Figure 6.3e.

We have said that the function of the dendrites is to pass information *towards* the cell body and that of the axon to convey it *from* the cell body. Although the action potential can run in both directions, in practice what happens is that a variety of signals arrive at the dendrites and set off a wave of electrical activity which passes through the cell body and to the axon. There, if the effect is large enough, it triggers an action potential which flows down the axon to the synapse.

The synapse is crucial to the workings of the nerve. Figure 6.4 shows the swelling at the end of the axon. Inside this swelling, apart from mitochondria and other organelles, are a number of small round structures, called vesicles, packed with a particular chemical. The arrival of the action potential at the synapse is the signal for the release of this chemical, which diffuses across the gap between the (*pre*synaptic) nerve cell and the (*post*synaptic) adjacent cell. The arrival of the chemical at the postsynaptic side is the signal for the second cell to begin its activities. If it is a muscle, it may begin to contract; if a gland, to secrete; if another nerve cell (as in Figure 6.2) an impulse is triggered in the second cell.

The chemicals which carry out these functions at the synapse are called neurotransmitters. The number of neurotransmitter substances is not known exactly (this is a growing area of research), but there are believed to be at least several dozen. Interestingly, several of them are chemically identical or very similar to hormones such as adrenalin and some of the pituitary hormones.

□ What does this suggest to you?
■ That they may have developed to their current specialised role from the hormones during evolution.

that small ions can either pass through or be trapped on one side of the membrane or the other. When the cell is 'at rest' (that is, not transmitting an electrical impulse) positively charged sodium ions tend to cluster on the outside surface of the membrane and negative ions cluster on the inside of the membrane. If you measured the electrical difference (voltage) between the two sides of the membrane, you would find that the inside was about 70 millivolts more negative than the outside. This is the so-called resting membrane potential, which all cells possess (Figure 6.3a).

What makes the neuron unique is that if a sufficient mechanical or electrical stimulus is applied to it somewhere along its length, the membrane changes its structure at that point. Sodium pours into the cell and the charge inside the cell changes until it is up to 400 millivolts positive with respect to the outside (Figure 6.3b). This 'charge change' is only transitory. As soon as the change has occurred, further entry of sodium is slowed at the point of stimulus and the *status quo* is restored. But the impulse, called the *action potential*, has begun! The charge change at the point of stimulus triggers the same process in the adjacent areas of the neuron and the impulse thus spreads further and further

Colour Plates

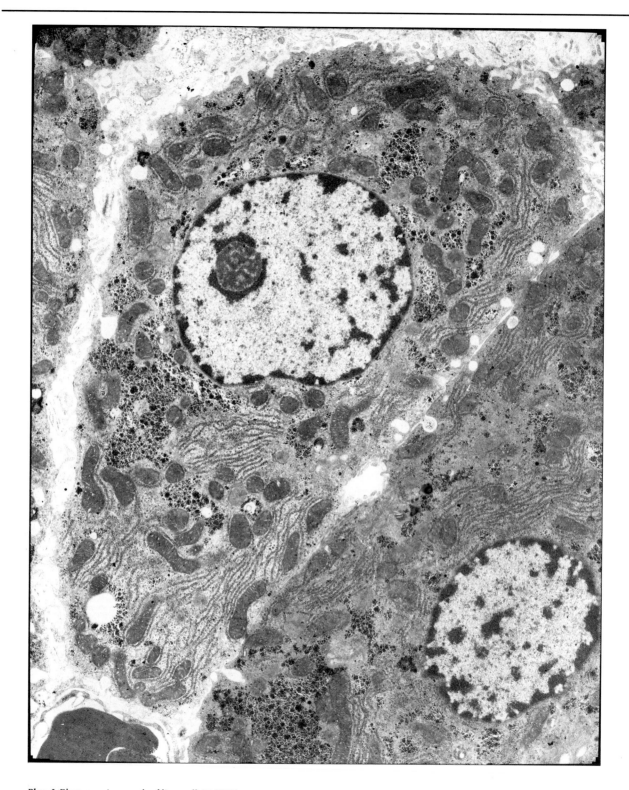

Plate I Electron micrograph of liver cell (× 8 000).

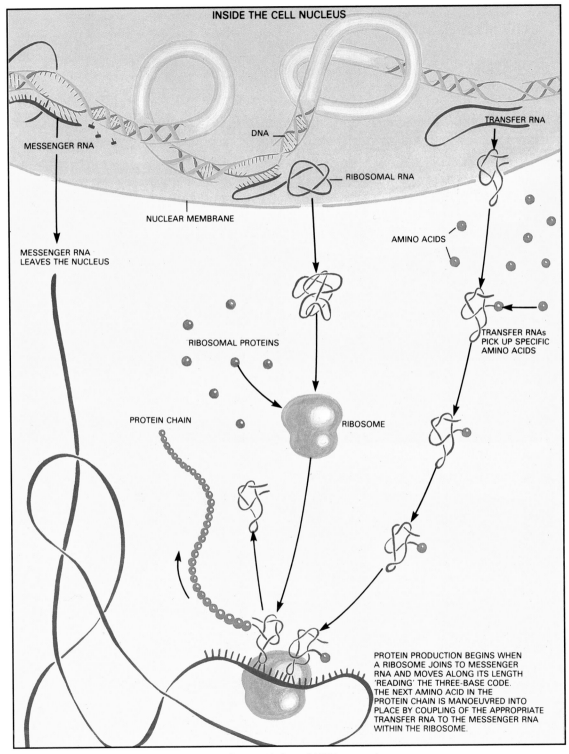

Plate II Protein synthesis: the flow of information from DNA through RNA to proteins.

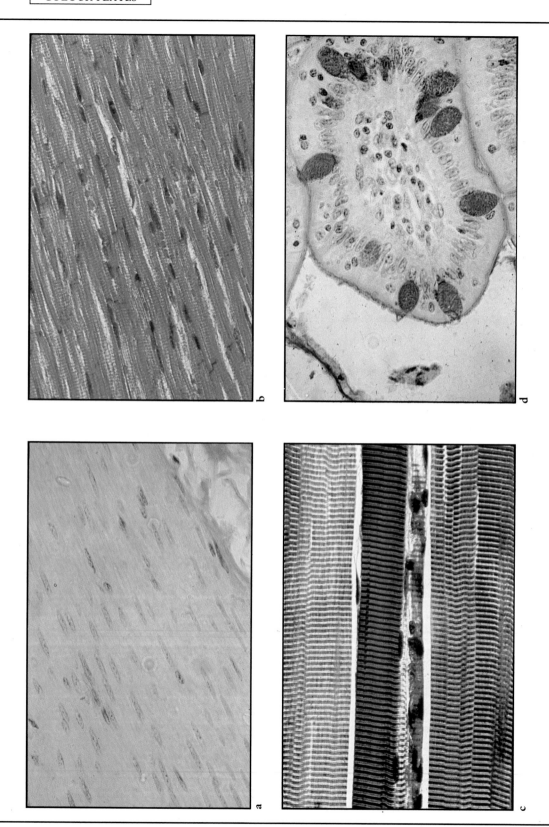

Plate III Sections of body tissues cut in 'bacon-slice' fashion and stained to show particular characteristics of the cells: (a) smooth muscle; (b) cardiac muscle; (c) striped (skeletal) muscle; (d) lining of the small intestine (the tube through which food passes is on the left-hand side). Bordering the tube are epithelial cells, some of which secrete enzymes and mucus (which appear green). Magnifications are approximately 500–750 times.

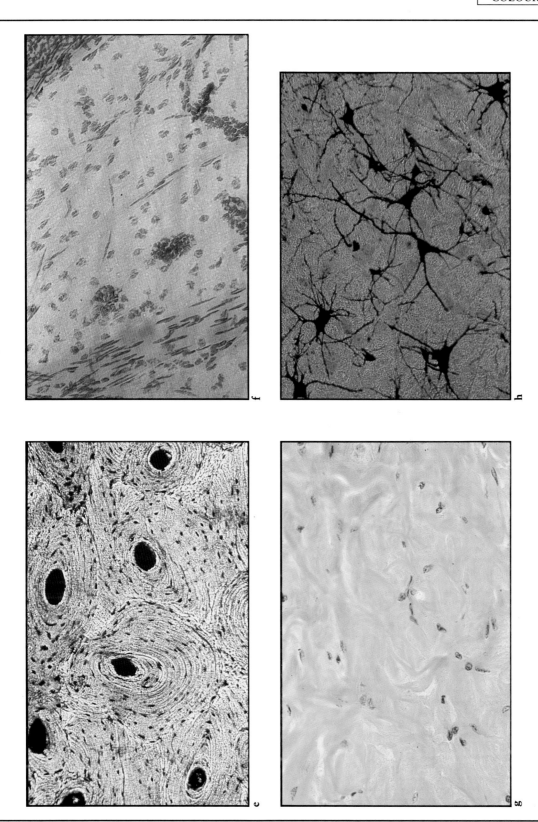

Plate III Sections of body tissues cut in 'bacon-slice' fashion and stained to show particular characteristics of the cells: (e) compact bone; (f) loose connective tissue; (g) dense connective tissue; (h) neurons. Magnifications are approximately 200–400 times.

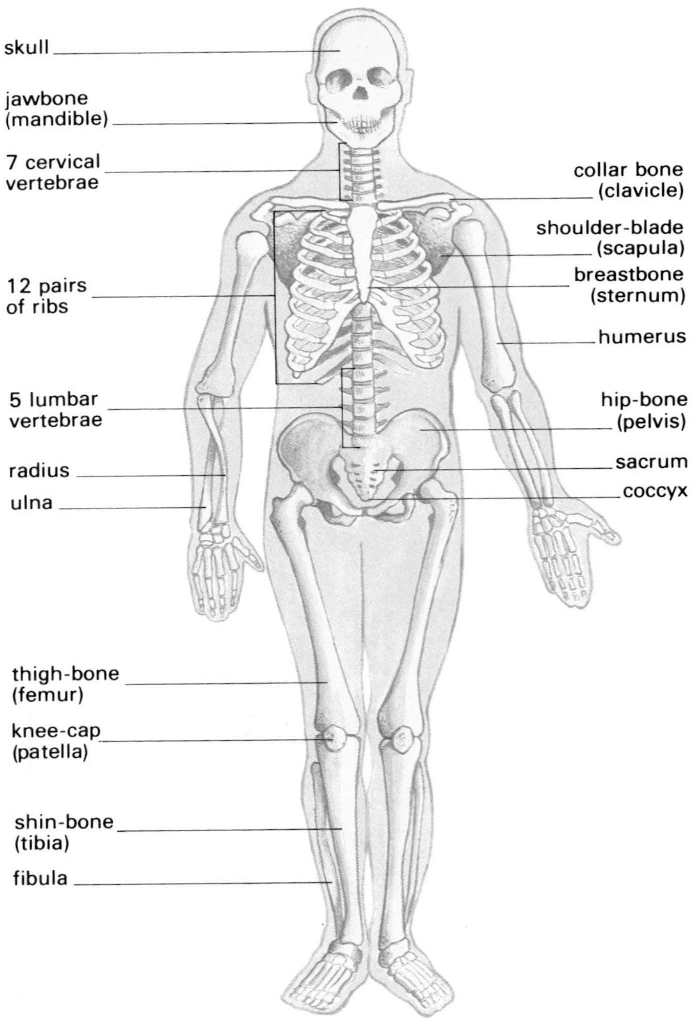

external
carotid artery

external
jugular vein

internal
jugular vein

aorta

pulmonary vein

heart

lower aorta

spleen

renal vein

kidney

femoral
artery

femoral
vein

brain

internal
carotid artery

pulmonary
artery

lung

hepatic
vein

hepatic
artery

liver

inferior
vena cava

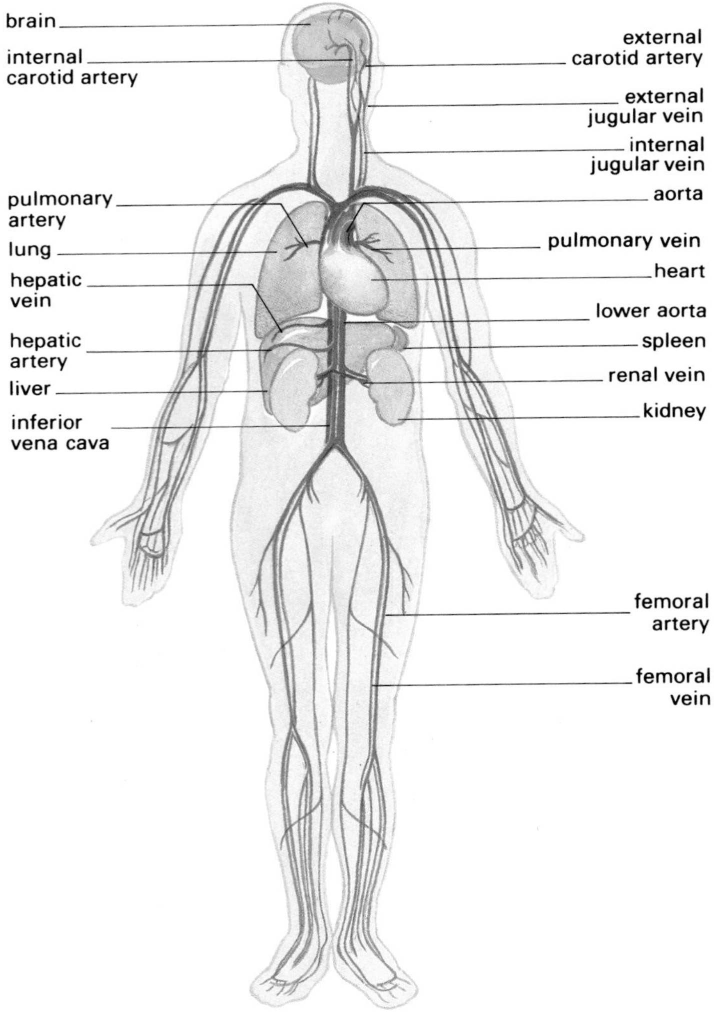

Plate V Circulatory system.

collar bone
(clavicle)

shoulder-blade
(scapula)

breastbone
(sternum)

humerus

hip-bone
(pelvis)

sacrum

coccyx

skull

jawbone
(mandible)

7 cervical
vertebrae

12 pairs
of ribs

5 lumbar
vertebrae

radius

ulna

thigh-bone
(femur)

knee-cap
(patella)

shin-bone
(tibia)

fibula

Plate IV Human skeleton.

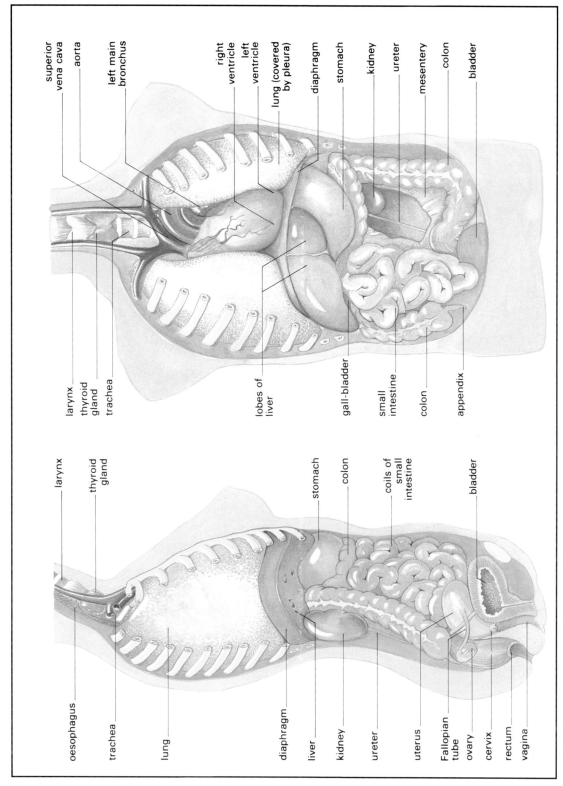

Plate VI Human torso: (a) side view and (b) front view showing arrangement of the major organs

Two other important points about synapses. The signal from the presynaptic to the postsynaptic cell may, as we said, be a positive one, to *do* something; the synapse and the neurotransmitter are said to be excitatory. The signal may, however, be a negative one to *stop* doing something and the synapse and the neurotransmitter are then said to be inhibitory. Thus the signals that nerves send are of two kinds, excitatory and inhibitory; the entire balance of the nervous system's function is the result of this interplay. The second point is that the synapse and its chemical neurotransmitter is a particular point of vulnerability in the workings of the nervous system.

□ What do you think might be the effect of a substance which interfered with neurotransmission between a nerve cell and a muscle?
■ Either the muscle would not contract when the message arrived at the synapse, or, if it was a message to *cease* contraction, the muscle would remain contracted and not relax. Either way, the result is highly inconvenient or may even be fatal.

From nerves to nervous systems
We have described the action of a single neuron. Both central and peripheral nervous systems contain thousands of millions such cells (the human brain alone has some 10 000 000 000 neurons). We have discussed synapses as if each axon connects with only one other neuron. In fact, within the brain each neuron may make synaptic contact with as many as 100 000 other cells. Clearly the subtlety and complexity of the messages that can be conveyed among all these cells is enormous. The cells and their axons are not arranged randomly, of course, but are connected up in a rather precise way. Let us consider what the function of this precise wiring is.

□ Can you think of two possible origins of nerve signals, and also two types of destination?
■ The most obvious origins are: the body's so-called sensory receptors, which transmit their 'messages' (about heat, cold, touch, pain, taste, smell, sight, sound) via nerves to the brain; and the brain itself, which sends signals to a variety of different body components.
Destinations include: muscles (nerves from the brain and other parts of the nervous system terminate in muscles and transmit electrical signals to the muscles by transmitter substances); and the brain, which receives signals from the sensory receptors.

Note that we are using the word 'nerve' to mean the axons running from a group of neurons in the brain towards, for instance, a muscle. 'Nerves' in this sense are bundles of many thousands of axons running along parallel routes. For instance, each of the optic nerves, which connect the eyes to the brain, has about a million axons in it.

There is, of course, a danger. If a bundle of axons run in parallel, a message travelling down one will act as a signal and provoke a message in an adjacent one, like a 'crossed line' in a telephone system. Partly to avoid this difficulty, each axon in a 'nerve' is wrapped in a lipid sheath which serves as insulation, rather like that on an electric wire. The lipid of which the sheath is composed is called myelin, and one of the most distressing of disease categories is that class known as the 'demyelinating diseases' of which multiple sclerosis is one of the best known. In these diseases, a steady breakdown of the myelin sheaths of the nerves gradually impairs their function.

One of the reasons for the serious and seemingly irreversible effects of these diseases is due to a unique feature of neurons: once they have formed and found their place in the brain and elsewhere, they must last a lifetime. Neurons do not divide and are not replaced. Once a cell dies the space it occupies is filled by other types of cells, or dendrites or synapses from nearby neurons. Although they do have a *limited* capacity for self repair (a feature of what is called the plasticity of the nervous system), the cells in the human brain, once dead, are dead and gone. In terms of neuron number, humans go downhill steadily from just after birth to old age!

We can now propose a rudimentary 'model' of the way the nervous system works.

SIGNALS FROM OUTSIDE WORLD	input via sensory receptors →	BRAIN (thought)	output to body parts →

This simple scheme is deficient for several reasons. First, it coveniently omits a whole range of body processes that are largely independent of sensory inputs from the outside world, and that there are *internal* sensors which signal imbalances in internal body processes (such as blood pressure) to the brain. Such signals must be taken into account. The scheme also omits the fact that many sensory inputs do not result in direct obvious outputs, but remain in the brain as 'perceptions' and 'memories'. Indeed, they may be received but just not acted on: a fly buzzing above your head may not be swatted if it goes away reasonably quickly and will not be remembered for long. Further, the brain connects to the spinal cord, a complex mass of neurons which is threaded down through the backbone. Many nerve pathways are routed to and from the spinal cord directly. A more accurate scheme for the nervous system can be seen in Figure 6.5.

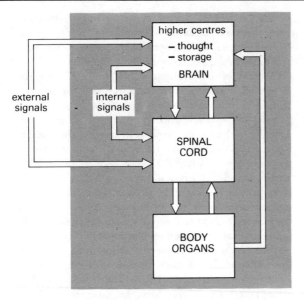

Figure 6.5 Schematic representation of the nervous system.

The autonomic nervous system

Before considering the central nervous system (brain and spinal cord) further, we should consider a 'lower' level of control of body systems much more akin to that provided by the hormones. To a very large extent, the day-to-day running of many body processes goes on without much direct central nervous system intervention at all. The digestive tract absorbs food material and solidifies waste. The bone marrow produces blood cells. The liver takes in, among other things, glucose and converts it to the carbohydrate glycogen. The body as a whole needs to exert control over these internal organs. We have already described some aspects of this type of control by way of the hormones, which regulate blood sugar level, heart rate and so forth. There is, however, an additional level of control that acts over and above these individual elements: the autonomic nervous system.

At times of emotional excitement or perceived danger, impulses travel down one set of nerves (called sympathetic nerves) to speed up the heart and provide it with more blood, at the expense of the skin and digestive organs. This is part of the 'fight or flight' reaction. This set of nerves also inhibits bladder contractions and stimulates the adrenal gland to produce the hormone, adrenalin. Adrenalin aids the re-distribution of blood from the internal organs to the skeletal muscles by constricting blood vessels in skin and internal organs while dilating those of the heart, muscles and lungs. It also promotes the conversion of glycogen back to glucose in the liver and reduces blood clotting time (in preparation for injuries).

☐ Summarise the effects of this first set of nerves.

■ The body is prepared for muscular activity and exertion. By moving blood from the internal organs, any strenuous activity will be helped by an increased blood supply to the parts of the body needing enhanced supplies of glucose and oxygen.

Interestingly, the neurotransmitter for these sympathetic nerves is a very similar chemical to the hormone adrenalin. Therefore, many organs that are targets of the sympathetic nervous system can be stimulated by both nerve impulses and adrenalin from the adrenal glands. *Sustained* stimulation of many of these target organs is achieved more effectively by adrenalin than by nerve stimulation.

The second set of nerves (parasympathetic) has the opposite effects to the sympathetic, although it tends to act in a more piecemeal fashion, affecting one organ at a time. It participates in aiding digestion, and slows the heartbeat and allows the bladder to contract normally. In general, this set of nerves acts to conserve and protect body resources. The body normally 'balances' the activity of the two sets of nerves and hormones, and there is a constant interplay between them. At times of stress or danger (driving a car in heavy traffic or running from a raging bull), the sympathetic system is dominant. At times of relaxation, the parasympathetic system is more prevalent.

The reflex arc and the spinal cord

Automatic muscular movements of the body are an important part of the body's defences. You blink automatically when an object gets too close to your eyes, cough when food 'goes down the wrong way' and rapidly withdraw any part of the body that comes in contact with a painful stimulus—for example, a drawing pin. These actions are termed *reflexes*. They happen so fast that you have no time to take a conscious decision to move the muscles involved in the action, even though all of them are normally under conscious control: the hand, for example, is withdrawn from the pin a fraction of a second before the sensation of pain is consciously registered.

The clue to the speed of these reflex actions lies in Figure 6.6, which shows a circuit of nerves, known as a *reflex arc*. Reflex arcs relay information from sense receptors, such as those in the skin which register temperature, pressure and pain, to motor nerves in the spinal cord. These immediately stimulate the appropriate muscles to respond. The 'message' about the pin-prick doesn't have to travel up the spinal cord to the brain before the signal to remove the hand can be sent. In some reflex arcs, the sensory and motor nerves are directly connected to one another in the spinal cord, but may reflexes involve so-called interneurons shown in Figure 6.6, which spread the in-coming information from one sensory nerve to several motor nerves so that several muscles can be moved at once.

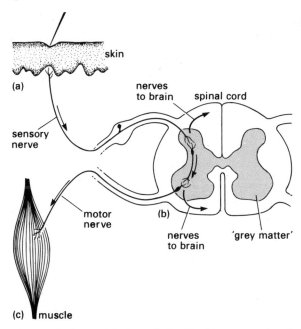

Figure 6.6 Reflex arc: (a) a sensor in the skin; (b) a cross-section of the spinal cord showing connection via an interneuron; (c) a motor neuron to a muscle.

Reflex arcs are also important in coordinating many conscious movements which nevertheless have an automatic component, for example, walking. Although the muscles of the legs and feet are under voluntary control, the rhythmic action of walking involves reflex arcs, which save you the trouble of making a conscious decision each time you pick up one foot and place it in front of the other.

With the exception of the nerves in the head, which run directly to and from the brain, most nerves lead to and from the spinal cord. Its importance as a relay centre, both of reflex responses and of information to and from the brain, can be seen in the results of spinal injury. If the cord is severed or damaged in an accident, severe loss of body function occurs, the severity depending on the site and extent of the damage. If it is cut at the neck, a total paralysis of nervous control below the site occurs as the cut nerve cells cannot reform their connections. This may include loss of respiratory function, as the nerves that control breathing lose their connection to the brain. Cuts lower down the cord are relatively less severe in their effects. Damage around the 'small' of the back may produce loss of function only below the waist.

The brain

The human brain is a dense mass of nerve cells, some without myelin sheaths—the 'grey matter'—and some with myelin — the 'white matter'. The number of synapses in the brain has been estimated at some 100 million million. In addition, there is supporting tissue and the adult human brain weighs about 1.5 kilograms.

☐ Make a list of the general functions with which you think the human brain is involved.

■ Thought, memory, communication with the outside world, perception of sensory information from the world, control of body activities, and emotion.

The organisation of this mass of cells is itself both ordered and very complex, the result of a process of evolution which has provided humans with a bigger brain for their body weight than any other animal except dolphins.

As you will see from Figure 6.7, viewed from the side, the spinal cord runs up into a thickened area called the brain stem above which are the large areas of cerebellum and cerebral cortex. Viewed from above (Figure 6.8), the brain is *bilaterally symmetrical*: that is, it is composed of two apparently identical halves. The functions of these halves, especially of the cerebral cortex, are not *quite* identical, though. The two halves are linked by a mass of myelinated nerves, the corpus callosum.

How does the brain's structure relate to function? Activities like breathing, swallowing, digestion and heartbeat are controlled from the medulla, as is the autonomic nervous system discussed above. Slightly higher in the brain is the cerebellum, which controls body movement and balance. Numerous nerve fibres pass from the medulla through various switching systems (notably the thalamus, to which most sense organs connect) up to the 'highest' part of the brain, the cerebral cortex. Below the thalamus is the hypothalmus, which regulates hormone activity and processes like temperature control.

The cerebral cortex is concerned with all higher mental processes including the registering of sensations, the initiation of voluntary activities, thought, learning and memory. Although many of these processes are shared between different cerebral regions, it is possible to locate a 'visual area', an 'auditory area', a 'body sense area', and an area responsible for controlling movements (Figure 6.8). Because damage as a result of blood clots or accident to other specific areas of the cortex lead to defects in processes like language, problem solving, the ability to discriminate different forms and memory, these areas are thought to be directly concerned with these processes.

Although it is conventional to speak of centres or regions of the brain which control particular activities, the question of localisation is not as simple as this, any more than one could say that the volume of a radio is 'in' the volume knob, or the loudspeaker, or the connections between. All parts are involved, though some are more directly concerned with one function than another. In addition, in part because of the way it has evolved, the brain

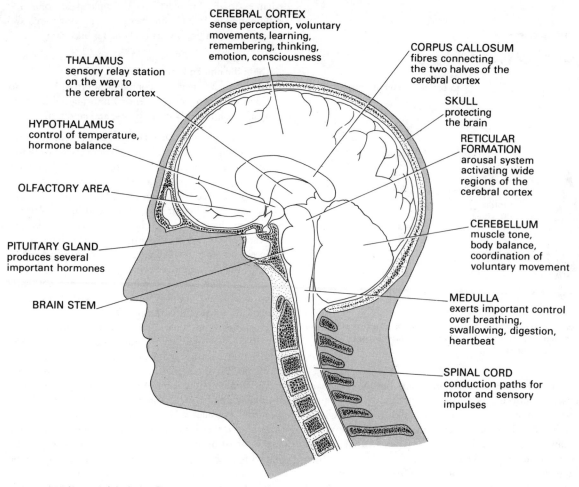

THALAMUS
sensory relay station
on the way to
the cerebral cortex

CEREBRAL CORTEX
sense perception, voluntary
movements, learning,
remembering, thinking,
emotion, consciousness

CORPUS CALLOSUM
fibres connecting
the two halves of the
cerebral cortex

SKULL
protecting
the brain

HYPOTHALAMUS
control of temperature,
hormone balance

RETICULAR
FORMATION
arousal system
activating wide
regions of the
cerebral cortex

OLFACTORY AREA

CEREBELLUM
muscle tone,
body balance,
coordination of
voluntary movement

PITUITARY GLAND
produces several
important hormones

BRAIN STEM

MEDULLA
exerts important control
over breathing,
swallowing, digestion,
heartbeat

SPINAL CORD
conduction paths for
motor and sensory
impulses

Figure 6.7 Side view of the brain showing the main structures and their functions.

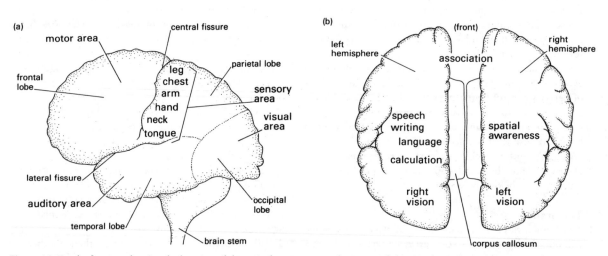

(a)

motor area

central fissure

parietal lobe

frontal
lobe

leg
chest
arm
hand
neck
tongue

sensory
area

visual
area

lateral fissure

auditory area

temporal lobe

occipital
lobe

brain stem

(b)

(front)

left
hemisphere

association

right
hemisphere

speech
writing
language
calculation

spatial
awareness

right
vision

left
vision

corpus callosum

Figure 6.8 Cerebral cortex showing the location of the main functions: (a) side view, and (b) view from above.

shows a characteristic redundancy of function, with particular properties apparently being duplicated in a variety of sites. Although neurons, once dead, cannot be replaced, during brain development and in response to damage new synapses and nerve pathways can be formed to take on new functions or as part of the process of learning new skills. This is an aspect of nervous system plasticity.

Sense organs

By 'senses' we usually refer to vision, hearing, taste, smell, touch, balance, pain and temperature. Although the modes of action of all these sensory processes are different, they share important common features.

□ What do you think these features might be?

■ They all input information from the outside world, and connect to the brain or spinal cord via neurons. All 'work' by the initiation and transmission of nerve impulses.

In some instances, the receptors (that is, the cells, or groups of cells, that 'transform' a stimulus into the electrical signal) are very simple. Pain and touch receptors are simply free nerve endings in the skin, in which impulses are initiated by physical pressure or injury. 'Smell' and 'taste' receptors depend on chemical changes in receptor cells due to minute quantities of the stimulant substance being absorbed by them.

The taste receptors are found in the taste buds on the edges and towards the back of the tongue, with a few located elsewhere in the mouth and throat. The receptors for smell are, as expected, found in the nose. Eddy currents of air carry smells to these cells, which are much more sensitive than taste buds, and the sense of smell is helped by the mucus-covered hairs on the exposed surface of the cells.

The ear works by mechanical transmission of sound vibrations via eardrum and 'amplifying' lever bones to a fluid-filled chamber in which minute hairs code the vibration into electrical impulses. The balance receptors are fluid-filled tubes lined with sensitive hairs that detect changes in orientation and are situated in the inner parts of the ear.

The eye is the most complex of the sense organs, conveying information about shapes, colours, textures, distance and depth to the brain. The human eye can be considered as an incredibly sensitive stereo colour camera, capable of generating very high resolution images. Instead of a light-sensitive photographic film, the visual image is projected onto the retina, a layer of specialised neurons which contain chemicals called photopigments. Changes in these pigments induced by the visual image on the retina generate impulses which pass from the retinal nerve cells across the retina and out of the eye via the optic nerve. They are then transmitted via a series of staging posts to the visual areas of the cerebral cortex. There they are processed and analysed by an ordered array of neurons which break down and then reconstruct the visual images.

We have discussed the workings of a variety of hormonal and nervous mechanisms for controlling body functions. We have also discussed the ways in which nerves can receive information from the outside world and transport it to the brain, where it is processed, stored, compared with past information and finally acted upon. The brain also has regions which regulate the release of hormones via the pituitary gland. These hormones are themselves affected by other hormones arising by way of the circulatory blood system. Other brain regions and networks of nerve cells are concerned with such phenomena as levels of awareness, and emotions such as anger, fear or sexual arousal. What about the properties most humans associate with brains — those of thought, consciousness and self-awareness, personal psychology and sense of life history? This is dangerously speculative ground, for philosophers, neuro-scientists and even psychologists are divided on these questions. It seems reasonable to argue that there must be some systematic relationship between processes that are described in the language of 'mind' as thought and consciousness, and 'brain' processes described in terms of the properties of nerve cells and their synaptic interactions. The exact details of this relationship can only be a theme for future study.

Objectives for Chapter 6

When you have studied this chapter, you should be able to:

6.1 Discuss the general contribution of the hormonal system to the regulation of physiological functions in the human body and show how this complements regulation by the nervous system.

6.2 Illustrate the range of effects mediated by hormones from the pituitary gland, and show how this gland directly or indirectly regulates metabolism in the human body.

6.3 Explain how the nervous system possesses a range of methods of control to provide an appropriate response, including reflex arcs, the autonomic nervous system, the brain stem and the cerebral cortex.

Questions for Chapter 6

1 (*Objective 6.1*) Which of the following statements about hormones are true and which false? Correct those statements that you think are false.

(a) Hormones are chemical substances that affect virtually all cells in the body by altering their rate of utilisation of food for energy.

(b) Hormones are released at a more or less constant rate in the healthy person, maintaining a sustained effect over a long time period.

(c) Hormones act more slowly than the nervous system, taking minutes, hours, days or even years before their full effect on the body becomes apparent.

(d) Some hormones are very similar to some neurotransmitters used by the autonomic nervous system and can provide sustained stimulation of many target organs more effectively than could be achieved by separate nervous stimulation.

(e) The output of hormones from the endocrine glands ceases when stimulation of these glands by the nervous system stops.

2 (*Objective 6.2*) Suppose that a juvenile mammal (say a laboratory rat) was given repeated daily injections of growth hormone, each dose containing significantly more hormone than the normal daily output of the pituitary gland. What effect would you expect this to have on the development of the rat, and what effect on the pituitary gland?

3 (*Objective 6.3*) Which part of the nervous system is responsible for each of the following activities?

(a) Increasing the heart rate during a frightening experience.

(b) Regulation of respiration.

(c) Recognition of visual images.

(d) Withdrawing a hand from a hot pan.

7

Transport

The changes that take place in the circulatory system at and immediately after birth are illustrated in the television programme *Life Before Birth*.

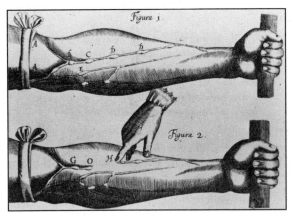

Illustration from the first edition of William Harvey's *De Motu Cordis* (1628) showing how blood in veins and arteries can flow in only one direction.

Many times in this book we have mentioned the basic needs of cells—the supply of nutrients, ions and oxygen, and the removal of the waste products of metabolism, including carbon dioxide. In this chapter we turn our attention to the transport system that fulfils these functions in the human body.

All the molecules needing transport to or from cells are small enough to pass through cell membranes, either by diffusion from regions of high concentration to lower concentration, or by being actively shunted across membranes by chemical 'pumps'. Equally important is the fact that most of these vital molecules are soluble in water and hence can be transported round the body dissolved in fluids.

All chemical reactions in the body, both inside and outside cells, take place in fluids. Recall from Chapter 2 that about 60 per cent of an adult human's body weight consists of water in which numerous molecules — proteins, sugars, fats, ions, vitamins and so forth — are dissolved. At any one time, about two-thirds of this fluid is inside cells (*intracellular fluid*). Some of the remainder exists in the spaces that occur between cells throughout the body (*extracellular fluid*). This space makes up 10–20 per cent of the total volume of all tissues. There is a continuous two-way traffic of fluids across cell membranes, as dissolved nutrients and oxygen are absorbed and soluble waste products expelled. How is the fluid bathing each cell kept rich in nutrients and low in toxic waste?

The answer lies in the fact that it is continuously being flushed away and replenished by fluids emerging from, and returning to, two major networks of tubes running to all parts of the body. The first network is the vascular system, made up of the blood vessels (Plate V), and the second, equally extensive, network of tubes is called the lymphatic system. They have rather different functions (discussed later) but for now you should be aware of two basic similarities. First, the fluid they contain is very similar in composition even though it has been given a different name in the two networks—plasma in the vascular system, and lymph in the lymphatic system. Second, both networks contain huge numbers of cells. The term blood refers to plasma together with the cells it transports. A cubic millimetre of plasma contains about 5 000 000 red blood cells in a healthy adult and about 7 000 white blood cells of several different types. One of these 'families' of white blood cells (the lymphocytes) also circulates in the lymphatic system and is a vital part of the immune response against infection (discussed more fully in Chapter 9).

The heart as a pump

Every time a person moves, the contraction and relaxation of muscles help to pump fluids around the body and through the tissue spaces, but in the vascular system the principal pumping mechanism is the heart. There are about 5 litres of blood in the average adult's circulation and, even at rest, the heart pumps at least this much through its chambers *every minute*.

The human heart has four chambers: two atria (singular, atrium) at the top, into which blood returns from the lungs and the rest of the body, and two much larger ventricles at the bottom which pump blood back around the system (Figure 7.1). The two atria beat simultaneously and force blood downwards into the ventricles. They then relax and the two ventricles beat together. This is why the heart has a two-phase beat. Each heart chamber contains valves which prevent reverse blood flow. For example, when the right ventricle contracts, the opening leading back into the right atrium is closed by blood pushing up the flaps of a 'non-return' valve. The right ventricle pumps blood to the lungs to collect oxygen, and the left ventricle pumps oxygenated blood to all other parts of the body.

The resting heart rate for an adult is around 70–80 beats per minute, or forty *million* beats per year! At each contraction, blood is forced out at high pressure into

arteries, vessels which have strong, muscular walls. The blood returns to the heart at much lower pressure after its long journey in much thinner walled veins.

☐ What keeps the heart beating? (You may wish to refer back to Chapter 3.)
■ The heart is constructed from specially adapted muscle cells (called cardiac muscle cells), which contract and relax *automatically*.

All the individual muscle cells, however, have to be *synchronised* so that they beat together. Figure 7.2 shows how this is achieved. A tiny bundle (or node) of muscle cells in the wall of the right atrium generates regular bursts of electrical activity, which spread through the two atria, causing them to beat rhythmically. This primary node acts as a sort of 'pacemaker' because it also transmits the signal to a secondary node in the right ventricle, which (after a brief delay) relays the signal to all parts of both ventricles, causing them to contract a fraction of a second after the atria.

☐ Why do you think it is necessary to synchronise the ventricles to contract a little *after* the atria?
■ Contraction of the atria fills the ventricles with blood. The delay before the ventricles contract ensures that they are completely full, and not 'wasting effort' by contracting on a partly filled chamber.

The heart has to perform sustained muscular *work* and must be 'fuelled' like any other muscle. Blood is supplied to the heart by the coronary arteries, which provide the heart muscle with the necessary oxygen and nutrients to fuel its

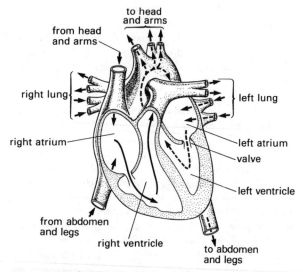

Figure 7.1 The circulation of blood through the heart. The circulation of deoxygenated blood is shown by a solid line and that of oxygenated blood by a broken line.

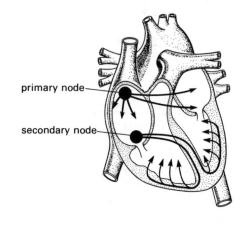

Figure 7.2 The route of the contraction wave in the heart, starting from the primary node (pacemaker) and spreading via the secondary node to the ventricles.

continual activity. Blockage of these arteries can rapidly lead to damage to the heart muscle.

The amount of blood pumped by the left ventricle of the heart into the main arteries to the body is termed cardiac output. As you might expect, it is tiny in a newborn baby, less than 2.0 litres per minute, but rises to about 5.0 litres per minute in an adult at rest, declining gradually during old age. A young, well-trained athlete may be able to pump as much as 30–35 litres of blood per minute during strenuous exertion.

Blood

Having considered how blood is pumped around the body, we now need to discuss the various essential functions it performs. Blood is made up of four main constituents:

1 A nearly colourless plasma, which carries dissolved nutrients obtained from the breakdown of food; toxic, waste products formed during biochemical reactions in cells, in particular carbon dioxide; chemicals which are involved in the clotting process; and hormones being delivered to their target tissues.

2 White blood cells which are involved in body defence processes, either by engulfing foreign material or by producing antibodies or toxic substances.

3 Red cells, coloured by the iron-containing protein haemoglobin which has a high affinity for oxygen. Red cells, formed in the bone marrow, lose their nuclei as they develop and are then released into the blood. Once there, they are concerned with the uptake and transport of oxygen to the body tissues. Haemoglobin is bright red when bound to oxygen, but dark purple in the absence of oxygen. Hence the different colour of blood in arteries and veins.

4 Platelets, which are tiny cell fragments without nuclei. They stick together at the site of any injury to blood vessel walls, forming a scab that helps to seal the breaks.

Curiously enough, *all* the cells in blood (including platelets) are derived from the same 'ancestral' cell. In the marrow of the long bones in the arms and legs, breastbone, vertebrae, ribs and pelvis, so-called multipotent stem cells are found. These cells divide repeatedly to give generation after generation of daughter cells, some of which eventually differentiate into red blood cells, some into the various types of white blood cell and some into cells that shed platelets into the bloodstream.

Turnover of cells within the blood is very rapid. Platelets live for only 2–10 days, red cells live somewhat longer (up to 120 days) and some kinds of white cells can last several years. Worn-out cells are broken down in the spleen (an organ that stores large quantities of blood cells and plasma) or in the liver, and new ones are continually being made in the bone marrow.

The blood cells make up about half the 5 litres of an adult's blood. A newly born baby weighing 3 kilograms has

as little as 0.2 litre of blood, and a pregnant woman will, shortly before giving birth, have up to 6.5 litres of blood.

The circulation of blood

The heart is at the junction of a *double circulation* as shown in Figure 7.3. One 'loop' pumps blood to and from the lungs, the pulmonary circulation ('pulmonary' is the adjective applied by physiologists to the lungs). The other 'loop' supplies blood to the rest of the body and then collects it again, the systemic circulation. At any one time, about 16 per cent of the total blood volume is in the circulation to or from the lungs, and about 84 per cent in circulation in the rest of the body.

☐ Can you see the advantages of this double circulation from Figure 7.3?

■ It ensures that *all* blood returning to the heart is pumped first to the lungs, to pick up oxygen and discharge carbon dioxide, before being distributed to the rest of the body. If there were a single circulation, only a portion of the blood leaving the heart would find its way to the lungs and some tissues in the body would receive blood that was still low in oxygen and high in carbon dioxide.

The left side of the heart receives blood rich in oxygen from the lungs, and pumps it out at high pressure to the rest of

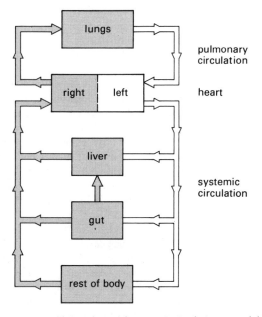

Figure 7.3 Double circulation: the systemic circulation around the body and the pulmonary circulation to and from the lungs.

the body. The right side of the heart receives blood low in oxygen (deoxygenated blood) from the body, and pumps it out to the lungs.

    □ Can you suggest why someone who has a hole in the wall separating the left and right sides of the heart feels very tired and breathless?

    ■ Oxygenated and deoxygenated blood from the two sides of the heart will mingle, reducing the amount of oxygen in the blood being pumped to the body. Overall, the supply of oxygen for breaking down glucose to release energy will be reduced and the person will have to breathe more often in an attempt to compensate.

At birth, all babies have such a 'hole in the heart', together with a blood vessel that bypasses most of the flow to the lungs. The hole rapidly seals and the bypass vessel shuts down when the first few breaths inflate the lungs. However, this sudden switch in the circulation is sometimes incomplete and gives rise to what are called 'blue babies'.

The vessels which transport blood around the body are specially constructed to fulfill rather different tasks. Cross-sections of an artery, a vein and a capillary are shown in Figure 7.4. We shall discuss each in turn.

Blood leaves the heart at high pressure, pumped in rhythmic pulses as the ventricles contract and force blood into the major arteries. *Blood pressure* (like atmospheric pressure) is measured in terms of the ability to support a column of mercury in a glass tube. At the maximum contraction of the heart in a healthy adult, blood pressure in the arteries rises to about 120 millimetres of mercury (that is, it can hold up a column of mercury 120 millimetres high). At the point of maximum relaxation arterial pressure falls to about 80 millimetres of mercury: you may have heard a nurse or doctor talking about normal blood pressure as '120 over 80'. The measurement at maximum contraction is called the systolic pressure and that at relaxation the diastolic pressure. Even in apparently healthy people, blood pressure shows an upward drift with age. The average of the systolic and diastolic pressures is around 70 millimetres of mercury in newborn babies, 100 mm in adults and 110–120 mm in elderly people.

The arteries branch into finer and finer tubules and eventually divide into minute vessels with very thin walls, called capillaries, each less than 1 millimetre in length and of even narrower diameter. By the time blood reaches these capillaries, it has a pressure of only about 30 millimetres of mercury, and the pulse is no longer detectable. Plasma seeps through the capillary walls and enters the spaces between cells, bringing fresh supplies of oxygen and nutrients. Almost no cell in the body is further than a fraction of a millimetre away from a capillary, so diffusion of dissolved substances takes no more than a few seconds. Even though capillaries are so tiny that they contain only about 5 per

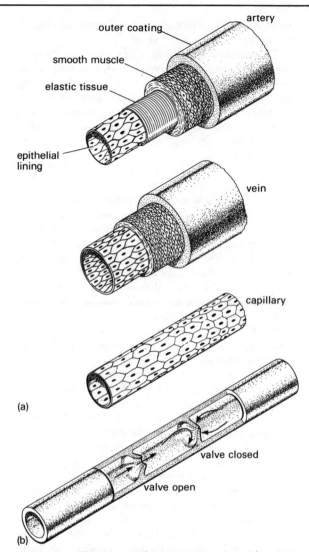

Figure 7.4 (a) The structure of arteries, veins and capillaries, and (b) valves in the veins preventing the reverse flow of blood.

cent of the total blood volume at any one time, their combined surface area is about 250 square metres (about the size of a tennis court), providing a huge surface for diffusion to and from the tissues.

Fluids containing dissolved waste products seep into other capillaries which gradually join to form the network of veins which return blood to the heart.

About 10 per cent of the fluid between cells returns by a different route—in the fine tubules of the lymphatic system. Molecules (for example, proteins) that are too large to pass through the walls of the blood capillaries drain into the open ends of lymphatic capillaries distributed throughout the tissues, and are eventually returned to the blood circulation through a duct that joins a major vein just behind the collar-bone on the left side of the neck. The lymphatic capillaries also collect the majority of fats absorbed from the digestive tract. Thus lymph, although similar in composition to plasma, has a higher concentration of protein and fats.

□ Can you see a problem in getting blood back to the heart via the veins of the vascular system?

■ By the time that blood enters the veins, it is at very low pressure and, for the most part, has to flow 'uphill', against the force of gravity.

To overcome this problem, veins have regularly spaced one-way valves (Figure 7.4). Once blood has moved forwards through a valve, it cannot flow back. In addition, as each segment of a vein (that is, the section between two adjacent valves) is distended with blood, the muscular walls of the vein give a squeeze which forces blood upwards through the next valve. Movement of muscles and limbs also helps to shunt blood back to the heart against the pull of gravity. Arteries do not need one-way valves because the flow inside them is almost always aided by gravity, and the pumping of the heart pushes the blood along.

The walls of all blood vessels are muscular and, by contracting or relaxing, can alter the diameter or the 'bore' of the tube. A few chemical substances produced in the body (including some hormones) stimulate constriction of blood vessels (vasoconstriction in physiological jargon), and so too does the activity of the sympathetic nervous system, which sends nerves to all blood vessels. When this stimulation ceases, the vessels relax and resistance to blood flow along those vessels decreases. In this way, the flow of blood to different parts of the body can be altered. For example, after a meal, blood is routed preferentially to the vessels close to the digestive tract to pick up nutrients, and there is a relative fall in blood flow to the main limb muscles. (This may partly explain the notion that muscle cramps will result if you go swimming after a meal.)

The state of constriction or dilation of blood vessels has a significant effect on blood pressure. More than 60 per cent of total blood volume is held in the veins, so constriction of the veins can have a profound effect on blood pressure.

□ What effect would a generalised state of vasoconstriction have on blood pressure?

■ It would rise, since the same volume of blood is being forced to travel in tubes of smaller diameter.

The regulation of blood pressure is enormously complex, involving sensors which monitor pressure by the degree of stretch in vessel walls. They also stimulate adjustments both in the rate and force of heartbeat, vasoconstriction and dilation and, more slowly, water retention by the kidneys. Notice that this regulatory mechanism has all the hallmarks of the 'systems view' of homeostasis discussed in Chapter 5—sensors, a feedback loop and adjustments to restore blood pressure to the normal 'set-point'.

□ Can you deduce another way in which blood pressure would increase, even if blood vessels were *not* constricted?

■ An increase in blood *volume* would cause an increase in pressure. (This happens when faulty kidneys make less urine and more fluid is retained in the blood.)

Objectives for Chapter 7

When you have studied this chapter, you should be able to:

7.1 Discuss the mechanisms that keep extracellular fluid circulating in the body, with special reference to the vascular and lymphatic systems, and the unique features of arteries and veins.

7.2 Demonstrate an understanding of the terms atrium, ventricle, cardiac output, heart rate and blood pressure by using them correctly in a discussion of the heart as a pump.

Questions for Chapter 7

1 (*Objective 7.1*) Answer (a)–(c), in no more than one or two sentences for each part of the question.

(a) Arteries carry blood away from the heart and veins return blood to the heart, but is it also true that arteries always carry oxygenated blood whereas veins always carry deoxygenated blood?

(b) If you hold your arm straight up in the air, your hand will begin to tingle within about 10 seconds as oxygen supplies become exhausted, and acid begins to build up in the muscles. Why don't adequate supplies of oxygenated blood reach a hand held in this position?

(c) Why do arteries have much thicker walls than veins, whereas veins have a much greater diameter than arteries?

2 (*Objective 7.2*) The rate at which the heart beats is partly determined by the rate at which the right side of the heart is filled by blood returning from the body. The faster the heart refills, the quicker it beats. What immediate effects on heart rate, blood pressure and cardiac output would you expect from (a) administering a drug that dilates veins, and (b) transfusing 1 litre of plasma rapidly into a major vein?

8
Regulating the internal environment

Some of the material on the regulation of blood glucose levels you will already have met in *Studying Health and Disease*. In this chapter the physiological mechanisms are discussed.

Claude Bernard in 1849.

In this chapter we return to the subject of homeostasis, or the regulation of the internal environment, as the French physiologist Claude Bernard described it when propounding the idea for the first time in the late nineteenth century. Within the body, a constant balancing act is going on as the conditions necessary to sustain life are monitored and adjustments are made to keep them as close to the optimum as possible. We could have chosen numerous different aspects of the internal environment to look more closely at how regulation is achieved, but we have settled on four of major importance: the regulation of oxygen levels in blood; the regulation of glucose levels in blood; the regulation of the water content of the body coupled with the excretion of soluble waste; and the regulation of body temperature. Some of the organs involved in these processes can be seen in Plate VI.

The regulation of oxygen levels in blood

You already know a great deal about the circulation of blood from the previous chapter, so it only remains to describe how the heart rate and respiration rate are adjusted to cope with fluctuations in the demand for oxygen. Begin by looking at the respiratory system illustrated in Figure 8.1. When the muscles covering the chest and those between the ribs contract, the ribs are raised and the diaphragm bulges downwards into the abdomen. The net effect is to increase the size of the chest cavity, thereby reducing the pressure around the lungs. Atmospheric pressure forces air to rush into the lungs through the nose or mouth and the windpipe. As the muscles relax, the chest 'collapses' and air is squeezed out again. Manual methods of artificial respiration (as opposed to 'mouth-to-mouth') utilise the same principle — by raising the patient's arms, the chest cavity is enlarged enough to suck in some air.

The windpipe (trachea) divides into two branches (each called a bronchus; plural, bronchi) carrying air to and from each lung. The bronchi divide into finer and finer tubules which eventually terminate in tiny air-filled bags called

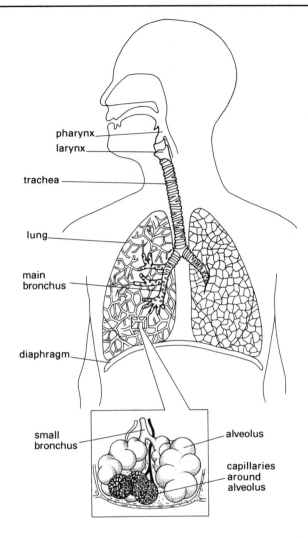

pharynx
larynx
trachea
lung
main
bronchus
diaphragm
small
bronchus
alveolus
capillaries
around
alveolus

Figure 8.1 The respiratory system.

alveoli. All lung tissue has a natural elastic tendency to collapse when not filled with air, so the real work of respiration is done during breathing in. The alveoli will inflate only if coated with a thin film of a secretion called surfactant; inadequate production of this lubricating fluid by premature babies leads to poor respiration. Blood vessels passing into the lungs also branch into a network of fine capillaries that cover the walls of the alveoli, allowing the exchange of oxygen and carbon dioxide.

☐ Why does carbon dioxide leave the capillaries and enter the alveoli, whereas oxygen travels in the opposite direction?

■ The respiratory gases, oxygen and carbon dioxide, diffuse along so-called 'concentration gradients'. Thus

they diffuse through the walls of the lung capillaries and alveoli from areas of high concentration to areas of lower concentration. Air has more oxygen and less carbon dioxide than the blood entering the lungs and so these gases travel in opposite directions.

Carbon dioxide which has diffused into the alveoli is expelled with the out-breath. At rest, about 0.5 litre of air is taken in at each breath by a healthy adult, but during strenuous exercise this volume can rise to about 3 litre. The concentration of carbon dioxide and oxygen in blood arriving at or leaving the lungs remains remarkably constant, no matter what the state of rest or exercise of the body. Obviously, if more oxygen is needed breathing is more rapid and deeper. How is respiration regulated so precisely that the supply of oxygen always matches demand?

☐ From your knowledge of the general features of mechanisms regulating homeostasis, what 'devices' would you expect to find in the control of respiration?

■ Sensing devices that sample the concentrations of oxygen and carbon dioxide in the blood, and compare these with the 'desired' levels (set-points); mechanisms that adjust the rate of respiration to push the actual levels towards the set-points; and a feedback loop that reduces respiration when the set-points have been reached.

There are two main sensing devices in the regulation of respiration. One, the respiratory centre in the brain stem, responds mainly to carbon dioxide levels. Oxygen concentration is detected by two small clumps of nerve tissue lying against the main arteries in the neck. If the carbon dioxide level rises, or the oxygen level falls (and they can shift independently of each other, as you will see in a moment), electrical impulses from the respiratory centre to the chest muscles and diaphragm stimulate increased respiration. As the concentration of dissolved gases in the blood approaches the desired set-point, stimulation falls back to the normal level. If, however, the demand for oxygen rises (say, during exercise), it is not enough simply to increase respiration. The supply of oxygenated blood to the tissues must also be increased, by increasing heart rate and cardiac output. If you have time, find your own wrist pulse by pressing three fingers of one hand against the inside of the other wrist, in line with the thumb on that side. Count how many times your pulse beats in one minute. Then, still holding your pulse, perform some mild exercise (stand up and sit down ten times). Notice how, soon after the exercise begins, a detectable rise occurs in pulse rate from around 70–80 beats per minute while sitting quietly to perhaps twice that during exercise. In really strenuous exertion, the heart may beat over 200 times a minute.

The main mechanism by which heart rate is adjusted to meet the fluctuating demand for oxygen is by electrical stimulation from the autonomic nervous system.

☐ Can you recall, from Chapter 6, the general effects of the sympathetic and parasympathetic branches of the autonomic nervous system, and hence predict what their effects will be on heart rate?

■ The sympathetic nervous system stimulates increased activity of numerous muscles and glands in the body, and hence stimulates an increase in heart rate and force of contraction. The parasympathetic stimulation has the opposite effect.

The sympathetic nervous system uses adrenalin as its neurotransmitter. If a sustained increase in heart rate is needed, the adrenal glands release adrenalin into the bloodstream, which produces a prolonged response.

Human beings live all over the Earth at altitudes up to about 5 000 metres above sea-level, so you can deduce that supplies of oxygen in the air are generally adequate. At sea-level, oxygen forms about 20 per cent of air, but with increasing altitude, the availability of oxygen decreases. People who are used to living at sea-level will notice the difference in oxygen pressure from about 2 500 metres, when they may feel unusually breathless on mild exertion. At 4 000 metres they may be breathless at rest and suffer from headaches and insomnia. A few people suffer from nausea, vomiting and giddiness, reactions that may settle after a day or so. Occasionally, increasing breathlessness and discomfort occur because the alveoli in their lungs become filled with fluid. The only sure cure for this is a return to low altitudes where the oxygen content of the air is higher. The degree to which mountain sickness occurs is related to the time taken to ascend. People who climb on foot suffer from few symptoms at first. By 6 000 metres, however, they become breathless. Above 7–8 000 metres, most mountain climbers use bottled oxygen.

Despite these difficulties, about 10 million people in the world live at altitudes above 4 000 metres, mostly in the Andes and Himalayas. They manage this by developing significant physiological adaptations, the most important being the increased concentration of haemoglobin in their blood. This may be up to one-and-a-half times the concentration found in people who live at sea-level, which means they can more easily pick up oxygen from the thin air. This increase in haemoglobin level occurs in anyone who goes up to altitudes above 2 500 metres, because low oxygen levels in the blood stimulate the bone marrow to form more red blood cells. As well as this compensatory increase in the oxygen-carrying capability of the blood, people at high altitudes experience changes in the arteries in their lungs and in the size of the right side of their heart.

Too much oxygen can also cause problems. Premature babies with breathing problems used to be given pure oxygen to breathe. It was later discovered that, when it was stopped and they started to breathe normal air, the drop in oxygen levels in the tissues at the back of the eye led, in some infants, to excessive growth of capillaries in the retina. This caused permanent blindness, a condition known as retrolental fibroplasia.

The regulation of blood sugar

People who fast to death usually die between about 30 and 60 days from onset of the fast, depending on their initial body weight, and assuming that they drink water. If they drink glucose dissolved in water, however, death occurs much more slowly, and it is many weeks before their muscles begin to 'waste'.

☐ Why?

■ Glucose is a major source of energy in the body. When fats and proteins are used as energy resources, they are first converted into carbohydrates which are then 'fed in' to the series of biochemical reactions that extract energy from glucose. Thus, if glucose levels are kept high, there is much less need to 'cannibalise' muscles for fats and proteins.

All the cells of the body have a constant need for a supply of energy. The fluids surrounding cells must therefore be kept constantly supplied with molecules that can be taken into the cells and oxidised to release energy. The most important of these molecules is glucose. Indeed, cells in the brain can use *only* glucose to supply their energy needs. We shall return in a moment to the question of how glucose is transported across cell membranes, but first we need to consider two other questions: how does the body acquire glucose from food, and how is the supply of glucose kept constant?

The breakdown of foodstuffs into glucose is just one of the functions of the digestive system (Figure 8.2). Other functions include the absorption of water and other nutrients such as amino acids, fats, vitamins, and minerals, and the excretion of some waste products from the liver. Before food can be absorbed it has to be broken down both physically and chemically into small molecules. Physical digestion is performed initially by the teeth and then by the stomach. The main purpose of physical digestion is to enlarge the surface area of the food for the digestive enzymes to act on.

Enzymes are released into the digestive tract at several points along the route taken by ingested food. The first site is in the mouth where the salivary glands secrete saliva containing, among other things, an enzyme that starts the chemical breakdown of starch. Other enzymes, which act on protein and fats, are released in the stomach. The site of greatest activity is the small intestine, into which enzymes control the level of acidity of the food. This is because the activity of all the enzymes is affected by the acidity level. In other words, each enzyme has an optimum level at which its activity is at a maximum.

☐ What other example of optimal conditions for enzymatic activity have you come across?
■ Temperature. In humans, the optimum temperature for enzymes is 37 °C.

Chemical digestion of food is completed in the small intestine. The small molecules that result can then be absorbed. The lining of the small intestine is covered in tiny projections, known as villi, which increase its surface area. Absorption occurs by means of diffusion and by active transport (if you are unsure of this term, then refer back to Chapter 2). The molecules of sugars, amino acids and other water-soluble nutrients are absorbed into blood capillaries in the villi. From there, the blood transports them directly to the liver, where they are processed further. In contrast, fats and fat-soluble nutrients (such as some vitamins) are absorbed into lymphatic capillaries which empty into the bloodstream, as described in Chapter 7. Finally, water is absorbed in the large intestine. Any remaining solid matter collects in the last part of the large intestine, the rectum, to await elimination from the body through the anus.

Returning now to the specific issue of glucose, we must consider how its supply is kept constant, given that humans do not eat constantly but tend to consume large amounts of food, two or three times a day. Except in people who suffer from diabetes mellitus, glucose levels in the bloodstream stay close to 4.5–5.0 millimoles per litre (equivalent to about 1 gram per litre) of blood. Even after a sugary meal the level rises no higher than about 8.0 millimoles per litre, and within two hours has returned to the set-point. In a well-nourished individual on a total fast, it is several days before blood glucose starts to drift much below normal. This suggests that there is a mechanism for storing glucose and then 'dribbling' it into the bloodstream at a rate that exactly matches the rate at which glucose is being used up by cells. This mechanism centres on the hormone insulin and the liver.

☐ Can you suggest how insulin and the liver regulate blood glucose levels?
■ When blood glucose levels rise after a meal, cells in the pancreas release insulin into the bloodstream. This causes excess glucose to be absorbed from the blood by the liver. Glucose is converted in the liver into a storage carbohydrate (glycogen) which is gradually reconverted to glucose and returned to the blood as insulin levels fall.*

* This is dealt with in *Studying Health and Disease*. The Open University (1985) *Studying Health and Disease*, The Open University Press (U205 *Health and Disease*, Book I).

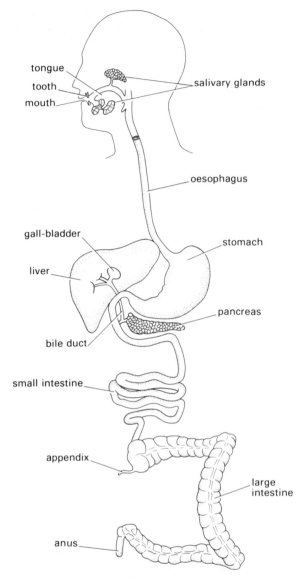

Figure 8.2 The digestive system. (The 'tube' has been straightened out somewhat.)

are secreted from the liver and pancreas. Enzymes are also an integral part of the intestinal wall.

The liver secretes a fluid known as bile, which is concentrated in the gall-bladder before passing into the small intestine. Bile contains essential enzymes for digestion, and waste products which must be eliminated from the body. Bile has a particularly important role in the digestion and subsequent absorption of fats. Pancreatic juice contains several enzymes for the digestion of proteins, nucleic acids, fats and carbohydrates. In addition to secreting enzymes, each part of the digestive system has to

The 'sensors' in this homeostatic mechanism are predominantly in the pancreas, in specialised cells which make insulin. These cells 'dump' their contents into the bloodstream within a few minutes of glucose levels rising in nearby capillaries. They then increase their rate of insulin synthesis and secretion. When glucose levels fall sufficiently, they simply stop secreting insulin. These cells are also sensitive to some of the digestive enzymes secreted by the stomach and intestines. When the level of these enzymes rises in the bloodstream, they trigger off an 'anticipatory' release of insulin, ready and waiting for the burst of glucose soon to follow. When blood glucose levels fall back to normal, insulin release stops and this is enough in itself to trigger the gradual reconversion of glycogen to glucose in the liver. However, in times of sudden need for a large increase in glucose (for example, during exercise) another hormone, called glucagon, is released by other cells in the pancreas and this speeds up the reconversion. During sudden stress, adrenalin from the adrenal glands also has the same effect.

If blood glucose falls below a certain level, this is detected by the so-called 'feeding centre' in the hypothalamus. The body experiences the complex sensation known as hunger and food is eaten. When glucose levels return to normal, another part of the hypothalamus, known as the 'satiety centre', is stimulated and eating is stopped. Of course, eating is much more than an automatic reflex and is capable of modification from the 'higher' conscious parts of the brain, but there is no space to discuss such complexities here. However, you should notice that the regulation of blood sugar involves the *behaviour* of the whole organism, as well as involuntary physiological mechanisms.

Finally, we must consider how glucose is transported across cell membranes. In the brain, glucose simply diffuses passively through nerve cell membranes, from the higher concentration in blood and tissue fluids to the lower concentration inside the cells. In other cells, glucose must be pumped across the membrane attached to a carrier molecule, which can be activated only by insulin. This explains why, in diabetes mellitus, glucose levels can get as high as 70 millimoles per litre of blood, about fifteen times the normal concentration. The lack of or insensitivity to insulin in this disease prevents both the liver from storing glucose in the form of glycogen, and cells outside the brain from absorbing glucose.

□ Normally the concentration of glucose in the fluids around cells is the same as that in blood. What do you think will be the effect if a cell is surrounded by a very concentrated glucose solution that cannot get in?

■ As glucose cannot diffuse from outside the cell (an area of high concentration) to within the cell (low

concentration), water will diffuse in the opposite direction. Thus cells become dehydrated in diabetes, as water floods into the tissue spaces, and ultimately the bloodstream, to dilute the excess glucose.

We have not yet discussed the role of the kidneys in regulating the concentration of sugar in the blood. When glucose levels get too high, as in diabetes mellitus, the kidneys filter some of it out of the blood and into urine. The result is that large amounts of urine are produced, flushing away not only some of the excess sugar, but also valuable ions, in particular those necessary for maintaining the function of nerves and muscles. This illustrates how easily a disruption of normal kidney function can have widespread effects on metabolic processes and, if uncontrolled, can soon lead to coma and death.

□ If diabetics inject too much insulin into their bloodstream, they may begin to hallucinate, feel panic, convulse or even fall into a coma. How do you explain this?

■ Cells in the brain can use *only* glucose for energy (they cannot switch to fats like other cells). Therefore, if blood glucose levels are pushed artificially low by excess insulin, the function of the nervous system is severely disrupted.

The great difficulty of balancing dietary intake of glucose and injected insulin in diabetes mellitus illustrates the subtlety and efficiency of the homeostatic mechanisms that normally regulate blood sugar.

Water balance and excretion

Although solid, insoluble waste is expelled from the intestines without ever entering the bloodstream, many toxic molecules are produced either as a result of digestion of foodstuffs or as waste products of cell metabolism. Some are rendered harmless by enzymes in the liver as the blood passes through that capillary-rich organ. Most have to be expelled from the body because their rising concentration in the bloodstream would cause serious disruption to the fine balance of cell metabolism. The most important of these waste products is urea, from protein digestion.

The organs of excretion of soluble waste are the kidneys, situated just above the waist on either side of the spine. Each receives blood via a major artery, the renal artery (renal is the adjective applied by physiologists to the kidneys). When blood from this artery reaches the outer part of the kidney (the cortex, see Figure 8.3), it is passed under pressure through an extremely fine network of capillaries. Most of the excess liquid and small soluble molecules are forced out through the capillary walls but blood cells and larger molecules, such as proteins, are retained. The filtered fluid travels via collecting ducts towards the centre (medulla) of the kidney. As it does so, more than 99 per cent of the water is reabsorbed, together with several important substances such as glucose, amino acids and salts (predominantly sodium, potassium and chloride ions). The resulting urine is, therefore, a concentrated solution of waste substances. Every day the kidneys of an adult filter out of the blood about 180 litres of fluid (weighing more than twice the total body weight!) yet reabsorption of water by the kidney is so efficient that only 1–1.5 litres of urine are formed. However, in newborn babies the efficiency of the kidneys is very much lower than that of adults. Fluid is filtered out of the blood at less than half the rate measured in adult kidneys (allowing for the difference in adult and newborn blood volume) and relatively little of the filtered fluid is reabsorbed. Kidney function does not approach adult levels until the child is about one year old.

Urine leaves each kidney via a tube called the ureter and passes into the bladder where it remains until expelled from the body via a single tube, the urethra. Blood, cleaned of its waste but with its water content restored and important molecules still in solution, leaves the kidneys via the renal veins.

The kidneys are not only involved in the excretion of toxic molecules, however. Remember from the last section how they even filter out excesses of useful molecules like glucose. They also expel water whenever the body fluids become too dilute. The key to balancing how much urine is produced is antidiuretic hormone (ADH). Diuresis is the technical name for the production of lots of dilute urine, so you can tell from its name that ADH promotes the secretion of small amounts of concentrated urine. When ADH levels in the blood are high, more water is reabsorbed from the filtered fluid in the kidneys and returned to the blood; when ADH levels fall, less water is reabsorbed and hence more urine is produced.

◻ The amount of urine produced by the kidneys regulates the volume of another body fluid. Can you deduce what it is?
◼ Blood volume is regulated by the amount of water filtered out or reabsorbed by the kidneys.

Blood volume in adults is normally held constant at around 5 litres by the activity of the kidneys. When a large volume of liquid is drunk it is filtered out of the bloodstream very quickly, as any beer drinker will testify. However, babies cannot regulate their water balance so precisely and their tissues easily become swollen with excess fluid if they are given too much to drink. The 'sensors' which detect a rise in blood volume following an intake of fluids are nerve cells sensitive to the stretching of major arteries near the heart.

◻ When these receptors are stretched, what response do they eventually trigger in the pituitary gland?
◼ They cause a *reduction* in ADH secretion, so that urine output increases and reduces blood volume back to normal.

Excessive water intake is much less common than water loss. Under temperate conditions, the average adult at rest loses at least 1 litre of water a day in urine, sweat and expired air, more if they drink a lot. A minimum urine output of 0.4 litre a day is necessary to excrete salts and the waste products of metabolism. Humans, therefore, must drink at least 1 litre of fluid a day, more if it is hot or they are taking much exercise. In a very hot climate, up to 10 litres of sweat may be lost in a day and corresponding amounts must be drunk in order to 'balance the books' and keep blood volume constant. Babies cannot do this easily, and are vulnerable to dehydration in warm weather.

The concentration of waste products in the blood obviously rises when blood volume falls. Sensory cells in the hypothalamus detect this and stimulate an increase in ADH secretion, so that urine output also falls. A small region of the hypothalamus generates the complex sensation of thirst, a conscious desire for water. A person weighing about 70 kg contains approximately 40 litres of water and up to 2 litres of this can be lost without any visible

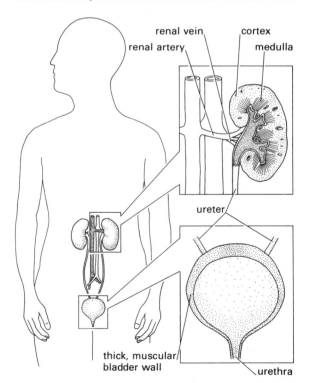

renal vein   cortex
renal artery   medulla

ureter

thick, muscular/
bladder wall
urethra

Figure 8.3 The urinary system.

change. With additional loss, the tongue becomes dry and the skin inelastic. After the loss of 5 litres, a person feels weak and may become confused. Death normally occurs after a loss of 6 litres, which is about 15 per cent of the body water and 10 per cent of the body weight. Exceptions do occur. There is one reported case of a Mexican who was lost without water in the Arizona desert for eight days. When found, he had lost 25 per cent of his body weight, his tongue was black and shrunken, and his skin was so dry and shrivelled that his cuts did not bleed at all. With adequate rehydration, he made a complete recovery.

Water is necessary, and it must be reasonably pure. People lost at sea die much more quickly if they drink sea-water than if they drink nothing at all. This is because sea-water contains a greater concentration of salt than even the most concentrated urine that humans can produce. In order to excrete the excess salt from the sea-water, extra water from the body has to be used to make urine. Drinking sea-water thus makes you *lose* more water than you take in.

Although humans have two kidneys, the body can function perfectly well with one. If one kidney is removed, cell divisions increase in the other so that within a year it has almost doubled in size to cope with the extra load. However, if both kidneys function poorly, toxic waste accumulates in the body, blood volume cannot be regulated and such severe metabolic disturbances occur that death follows within a few days or weeks. An artificial kidney machine may be used to clean the blood, by removing it from the body and passing it over a semipermeable cellophane membrane through which waste products are absorbed. The 'laundered' blood is then returned to the body, but this process has serious disadvantages, not least the inconvenience of spending several hours hooked up to the machine for several days per week. Most people in this situation hope for the chance of a kidney transplant.

The regulation of body temperature

It is usual to think of 'normal' body temperature as being 37 °C (98.4 °F), but human beings are not that precise! It may be helpful to borrow a metaphor from technology: humans seem to have a 'thermostat' set to a particular level. In some people, 'normal' temperature inside the body may be regulated at around 36 °C (97 °F); in others around 37.5 °C (99 °F). In many active children and in adults engaged in hard work or, indeed, in emotional crisis, the set-point may be higher—38.5 °C (101 °F). Whatever the individual 'normal' temperature, a naked human could sit indefinitely in an external temperature in the range 12–60 °C (55–140 °F) and maintain a constant core temperature, provided that the air were dry. Core temperature is measured inside the body, whereas surface temperature fluctuates with the surroundings.

Heat is lost from the body continuously. The body radiates heat and heat is conducted away by objects touched and by the surrounding air. When humans breathe out, urinate or defecate, they lose heat as well as waste products. More subtle is the heat lost by evaporation of water from the skin and lungs. This water (up to 0.6 litre per day in temperate climates) absorbs heat from the body to change it from a liquid to water vapour. Even without extremes of exercise or atmospheric temperature, a human loses enough heat from water evaporation *every hour* to heat 1 litre of water by 12–16 °C (55–65 °F). In a day, that is enough to heat more than 3 litres of water to boiling point! The insulation provided by body fat, clothes and hair is an important factor in reducing heat loss.

☐ What generates body heat in the first place?
■ Many of the chemical reactions inside cells generate heat and so does the friction of muscular activity.
☐ What are the responses you make (either consciously or involuntarily) when you feel too cold or too hot?
■ When you are cold, you are probably aware that your skin looks very pale and that you shiver involuntarily. You also make certain conscious actions to warm yourself, such as putting on more clothes. When you are too hot, you sweat involuntarily and your skin becomes flushed. You may take conscious action, such as removing clothes.

All these responses are summarised in a flow chart (Figure 8.4), a useful shorthand with which to represent physiological systems. You should bear in mind that it is a grossly simplified account which leaves out numerous subtle interactions. It *may* be possible to specify all the influences of one input on another within the circuitry of a domestic central heating system, but the human body is far more complex. Leaving that caveat aside, you should be able to interpret Figure 8.4 quite easily from experience of other regulatory mechanisms discussed in this chapter.

☐ Where is the 'thermostat' in this system?
■ Again, in the hypothalamus. Blood passing through this versatile part of the brain is 'sampled', its temperature compared with the normal set-point for that individual and signals sent to higher brain centres and other parts of the body to bring about readjustments.

When body temperature rises too high, the autonomic nervous system stimulates the secretion of sweat from glands in the skin, which increases the heat lost by evaporation. Autonomic nerves also trigger dilation of blood capillaries in the skin, so that more blood passes close to the body surface and more heat is lost by radiation. Conversely, when body temperature falls too low, sweat production ceases and autonomic nerves stimulate con-

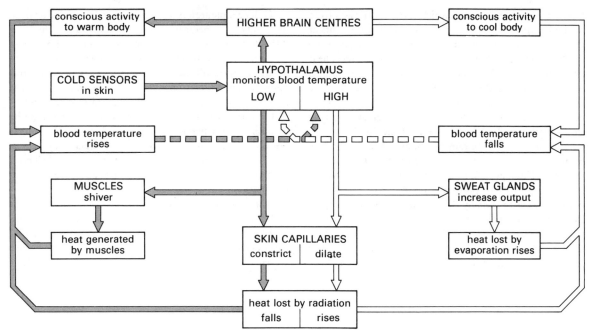

Figure 8.4 Regulation of body temperature represented as a flow chart. Grey arrows show responses to low temperature, white arrows to high temperature. The broken arrows show regulation of the response by negative feedback circuits.

striction of capillaries in the skin. The hypothalamus also generates electrical impulses that travel down the spinal cord and are relayed to skeletal muscles throughout the body via motor nerves. These impulses are irregular and change the 'tone' of the muscles so that they contract and relax in small spasms — shivering. During violent shivering, body heat production can rise four- or fivefold.

☐ Figure 8.4 shows that a *fall* in temperature is registered by cold sensors in the skin as well as by the hypothalamus monitoring blood heat. There is no such additional external sensor for *rising* temperature. Can you deduce why not?

■ Body heat is generated from *inside* the body, whereas cooling occurs from *outside*. Thus when the body heats up, this will be detected first by a sensor deep inside the body core, in contact with the bloodstream.

Another reason for needing a thermostat sensitive to heat in the body core is that body temperature may rise during illness, even though the conditions outside the body are cool. The most common causes of fever (elevated blood temperature) are the toxic waste products of certain infectious bacteria, or substances released by tissues damaged during viral infections. These pyrogens (substances that 'set on fire') greatly increase the heat sensitivity of the hypothalamus and (metaphorically) reset the thermostat to a higher temperature. A person will feel too hot or too cold only if there is a discrepancy between actual body temperature and the temperature set by the hypothalamus.

☐ Suppose that the hypothalamus is suddenly 'reset' to regulate blood temperature to 40 °C (103 °F). How will a person whose temperature is 37 °C (i.e. 'normal') experience this?

■ They will feel intensely cold, because actual temperature is lower than that set by the hypothalamus. This accounts for the chills and shivering commonly experienced by someone at the start of a fever, before their blood temperature has risen to the new set-point.

Once the blood temperature is stable at 40 °C, the person may feel unwell but will not experience uncomfortable coldness or overheating. When the concentration of the pyrogens falls (for example, when antibiotics take effect), the hypothalamus readjusts to the normal temperature, and the person (whose actual temperature is still 40 °C) temporarily feels extremely hot and sweats profusely until the blood temperature returns to normal. This is the fever 'crisis', a central motif in melodrama. Wise attendants see it as a good sign, heralding recovery. However, if body temperature rises much above 41 °C (106 °F), heat stroke is likely to occur. Irreparable damage is suffered by cells, especially in the brain, leading to dizziness, delirium, unconsciousness and ultimately death if the temperature remains high for long.

Uncontrolled loss of body temperature is equally

threatening to life. In water, which carries heat away from the body rapidly, humans can survive for about 6 hours at about 15 °C, but at temperatures close to freezing (0 °C) most people die within an hour. The water conducts heat away from the body faster than the body can produce it, with the result that the body temperature falls. By the time it drops to 30 °C, a person is usually unconscious. At 26 °C or thereabout, the heart beat becomes irregular, and may stop. Death usually occurs between 24 and 26 °C.

Humans usually maintain their body temperature in low environmental temperatures by means of insulating clothing or shelter. The importance of these two factors is demonstrated by the deaths from exposure that occur each year on the fells in the UK. Deaths usually occur in wet weather, since wet clothing has very poor insulating properties. Once the body temperature has dropped 3 or 4 °C, metabolism slows down to such an extent that brain function may be impaired. The resulting confusion may prevent people from reaching shelter.

Older people are less sensitive than young adults to a drop in body temperature. They may also find it difficult to move around to keep warm. Hypothermia (sustained low body temperature) is the recorded cause of death of about 100 elderly people every winter, but this figure is an underestimate, as hypothermia often goes unrecognised by doctors. For this reason, it is recommended that the homes of people in this age group should be maintained at a temperature above 18 °C (65 °F). Babies are also quite vulnerable to heat loss since they cannot regulate body temperature very efficiently, especially if born prematurely, hence the practice of putting such babies in an incubator.

This concludes the discussion of physiological regulation in the human body. In this chapter we have concentrated on four key aspects of homeostasis, but the general principles described here can be applied to the many other aspects that we had to omit.

Objectives for Chapter 8

When you have studied this chapter, you should be able to:

8.1 Discuss the contribution of conscious actions and involuntary regulatory mechanisms in maintaining homeostasis by reference to the control of oxygen and glucose levels in the blood, water balance and the excretion of soluble waste, and the control of body temperature.

8.2 Use examples drawn from this chapter to demonstrate that physiological regulation varies from one individual to another and during different parts of the life cycle.

Questions for Chapter 8

1 (*Objective 8.1*) Figure 8.5 represents the regulation of blood glucose levels as a 'flow-chart' constructed on the same principles as Figure 8.4, but with parts of several boxes left blank and most of the arrows between boxes omitted. Complete the diagram, then compare it with the version in the answers at the end of this book.

2 (*Objective 8.2*) Imagine that a baby and an elderly person were exposed to low atmospheric temperatures, say 5 °C (40 °F), for several hours but were given unlimited amounts of bearably hot liquid to drink. Assuming that the insulating effect of clothes and body fat was the same for each of them, what would be the consequences for their body temperature and urine output?

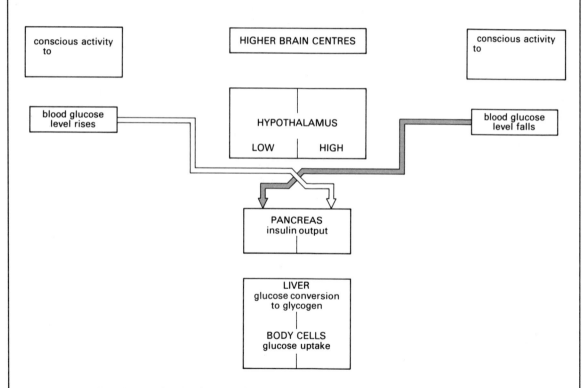

Figure 8.5 Flow chart showing the main features of mechanisms regulating blood glucose levels (incomplete).

9
Inflammation and immunity

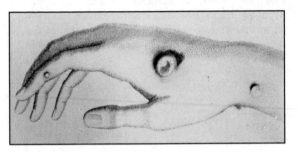

Hand of the milkmaid Sarah Nelmes infected with cowpox, from which Edward Jenner conceived the idea of vaccination.

However well the basic needs for food, oxygen and so forth are fulfilled, humans soon die if the physiological mechanisms for repairing injuries and keeping infections under control break down. In mammals, two interconnected reactions occur in response to injury and infection — the inflammatory response and the immune response.

Inflammation

Injuries and infections, from whatever cause, provoke inflammation within a few minutes around the damaged area. The *inflammatory response* varies little no matter where in the body the injury occurs, but it is difficult to extricate precise stages in the process since a cascade of biochemical and cellular events takes place almost simultaneously. The chemical histamine, and some other related substances, are of particular importance in triggering the inflammation.

Histamine is stored in the granules of specialised cells called mast cells (Figure 9.1). They are particularly abundant in the skin and mucous membranes of the respiratory and digestive tracts. Injury causes mast cells in the damaged area to expel their granules, producing a local surge of histamine that has profound effects on nearby blood vessels. They dilate, increasing the local blood supply, and their walls become 'leaky' so that plasma and white blood cells can flood through the vessel walls into the injured area. In a short time the injured site becomes packed with white blood cells of many different types.

□ What physical signs might you expect to follow such a change in blood vessel permeability and dilation?
■ A rush of fluids and cells into the area must result in local *swelling*; if the injured area is near the body surface, you would expect it to look *red* because the blood vessels are dilated; and, because the body surface is normally cooler than blood temperature, you would expect the inflamed area to feel *hot*.

These three signs (swelling, redness and heat), plus a sensation of pain and some loss of function in the affected

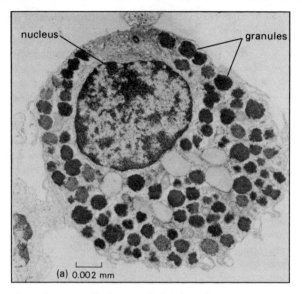

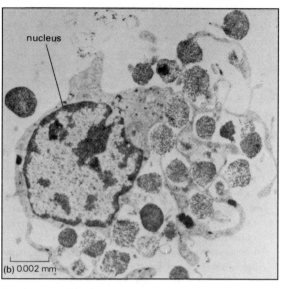

Figure 9.1 Mast cells photographed with an electron microscope. (a) The numerous dark granules inside this cell contain many chemicals (including histamine) which act on blood vessel walls.

(b) The mast cell has been triggered to release its granules. Despite the extensive disruption to the cell surface membrane as the granules pass through it, the cell recovers and manufactures more granules.

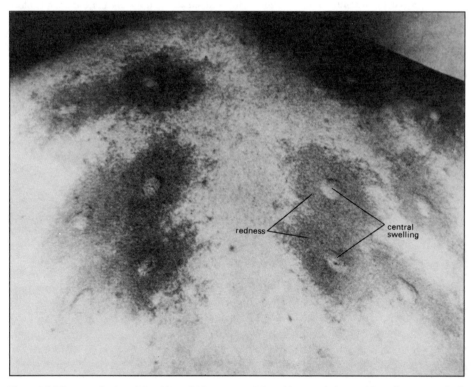

Figure 9.2 The upper back and shoulders of this man were injected in several places with a substance to which he is allergic. An acute inflammatory response developed within minutes around the injection sites. The characteristic 'wheal and flare' appearance may be more familiar from insect bites and nettle stings. (Courtesy of Dr D.R. Stanworth.)

area, are characteristic of an *acute* inflammatory response (see Figure 9.2). Inflammation does not (in the biological sense) mean bursting into flames or being provoked to anger, but it clearly takes its name from the 'fiery' or 'angry' quality of the physical sensation. Acute inflammation, as you would expect, arises within minutes, hours or at most a few days of contact with the injurious agent and usually subsides within a similar period.

☐ Can you list different categories of injurious and infectious agents that might provoke an inflammatory response?

■ You probably thought first of mechanical injuries such as grazes, burns from intense heat or cold, electric shocks and corrosive or irritant chemicals. Irradiation can also provoke inflammation, as anyone who has had radiotherapy will testify.

Infections by bacteria, viruses, certain fungi and helminths (such as intestinal worms or blood flukes) trigger inflammation in any area in which they settle. You may have thought of tissue decay following loss of blood supply to a part of the body, or as a result of invasion of normal tissues by a cancerous growth.

In addition, *allergies* are characterised by inflammation. For example, the nasal and bronchial passages are affected when the allergen (that is, the substance that provokes an allergic reaction) is airborne, like pollen grains. (As you will see later in this chapter, the immune system is the real 'culprit', causing injury to nearby tissue as it over-reacts to the allergen.) You may well be wondering whether inflammation is a 'good idea', since it is obviously painful

and to a limited extent damaging to tissues close to the actual injury.

☐ Can you deduce any biological 'advantages' of inflammation?

■ (i) Pain draws attention to the injury and may result in changed behaviour that promotes healing, such as keeping the affected part still.

(ii) Several types of white blood cell can engulf (phagocytose) debris, microorganisms and dead cells, thereby 'cleaning up' the area. Figure 9.3 shows one such type of cell, the macrophage (literally, 'big eater').

(iii) The influx of liquids may dilute toxic substances and literally flush them away.

Thus the inflammatory response is the first line of defence against infection, aimed at preventing potentially harmful organisms from establishing a viable 'colony' inside the body. Boils and pimples offer a useful example of how the inflammatory response works. Imagine that a small cut or puncture occurred in the skin and became infected with bacteria, or that bacteria began to multiply in an enlarged greasy skin pore. The presence of infection triggers mast cells to release their chemicals. Infected areas become swollen, hot and red as blood vessels dilate and white blood cells of the phagocytic types arrive in millions. Clear fluid derived from blood plasma may leak from the surface of the erupting spot, eventually turning to yellow pus coloured by bacteria and white blood cells in huge numbers. Tissues at the very centre of the inflammation may be killed by the reaction and gradually harden to form a scab, which seals the wound and prevents further infection. The inflamma-

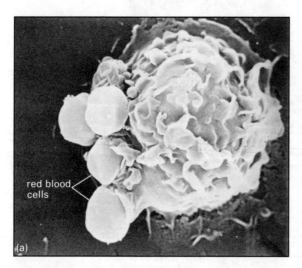
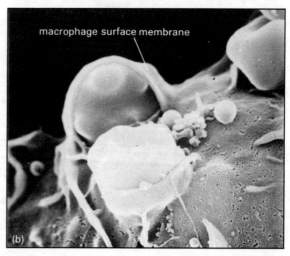

Figure 9.3 Macrophage photographed with a scanning electron microscope, which reveals the three-dimensional appearance of the surface of the cell. (a) Macrophage beginning to engulf red blood cells taken from a different species and, hence, recognised as non-self (magnification about 5 000 times). (b) Close-up showing the macrophage surface membrane folding over a red cell (magnification about 10 000 times).

tion is prevented from spreading far from the original lesion by specialised white blood cells on the perimeter of the response. They secrete enzymes that destroy histamine and the other chemicals that dilate blood vessels and make their walls 'leaky'. Within a short time the damaged tissues begin to regenerate: the incidence of cell division rises as new cells are formed to replace lost tissue and the superfluous white blood cells die or continue on their wandering existence.

An inflammatory response may sometimes persist for months or even years. *Chronic* inflammations generally lack the 'hot painfulness' of acute reactions, but show abnormal growth of connective tissues in the affected area (see Figure 9.4). Thus, raised ulcerating lesions may be seen, or substantial areas of scar tissue may develop, or harmful substances may become 'walled off' in a cyst. Chronic inflammation most commonly arises in response to chronic infection. For example, the bacteria causing leprosy, leprosy bacilli, often provoke such a severe chronic inflammation that the blood supply around infected areas is cut off, gangrene can set in and large areas of tissue may be damaged or destroyed. Untreated tuberculosis is accompanied by such chronic inflammation around clusters of tubercle bacilli that severe and eventually fatal damage to vital organs occurs. Multicellular parasites, such as worms and flukes, may also provoke chronic inflammation: in fact, this is our principal defence against them.

Immunity

Despite the effectiveness of inflammatory processes in confining infections and sealing wounds, harmful organisms may still penetrate deep into tissues or enter the bloodstream and become widely dispersed. A second line of defence on a *general* (rather than local) scale is provided by the *immune response*, which overlaps with the inflammatory response, both in its *function* (both are destructive to

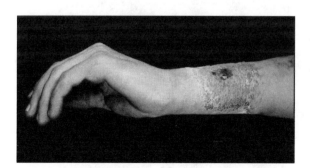

Figure 9.4 Chronic inflammation on the wrist of a person who is allergic to chemicals used in the tanning of leather. The response was triggered by a leather watch strap. Notice the scaly, raised appearance of the affected area, showing the abnormal growth of connective tissue that is characteristic of chronic inflammation.

pathogenic (disease-causing) *organisms*) and in its component *structures* (both depend crucially on the activity of white blood cells). The immune system, however, has a number of significant refinements, which we shall now examine.

The inflammatory response alone is not adequate to prevent the proliferation of many kinds of bacteria, viruses, fungi and other microorganisms, which can cause damage, either by producing poisonous waste, disrupting cells and tissues, or blocking vital organs and blood vessels. Later in this book you will learn much more about diseases that result from the effects of such pathogenic organisms, but for the moment we shall concentrate on the complex network of specialised cells and molecules that respond to these harmful visitors. This network is called the immune system and its activation by infectious organisms and parasites is termed the immune response.

☐ If you were attempting to design an immune system capable of protecting the body from the effects of harmful organisms and their products, what key features would you insist upon? (Think about the problems of accurate recognition that this must involve.)

■ The system would have to be capable of: (i) distinguishing between the cells and molecules of which the body is composed, called *self material*, and those of the organisms against which protection is required, *non-self* material*; (ii) inflicting damage selectively upon these non-self materials with little or no harm to nearby self tissues.

The simplest form of immunity that distinguishes between self and non-self is known as non-specific immunity. It comprises the inflammatory response *plus* a few additional chemical mechanisms aimed at microorganisms. For example, an enzyme known as lysozyme is found abundantly in mucus, tears, sweat and other secretions; it breaks down certain sugar molecules found *only* in bacterial cell walls. Another substance, interferon, is secreted by white blood cells and prevents virus particles from replicating. A third chemical mechanism is a sequence of enzymes in blood plasma (known collectively as complement) which can be triggered to digest holes in the membrane of non-self or infected self cells. Finally, it must be mentioned that macrophages secrete hydrogen peroxide (bleach!) onto bacteria, with unpleasant consequences for the bacteria.

* *Non-self* is a relatively recent term. Since the earliest study of the immune system, it has been fashionable to refer to non-self materials as *foreign*. However, the equation 'foreign = harmful' may have been acceptable in our more *chauvinist* past, but can no longer be defended!

However, these chemical mechanisms, plus macrophages and other phagocytic white blood cells, cannot tell one pathogenic organism from another, so it follows that they cannot 'remember' whether or not any given pathogen has been encountered before. Therefore the response is no more effective on the second or subsequent infection than it was on the first occasion when that organism entered the body.

□ Does this type of response fit your experience of infectious disease?

■ As you may have found, numerous infections hardly ever recur once you have recovered from them. All the so-called 'childhood infections' such as rubella (German measles) and chickenpox rarely arise in adults precisely because the first infection gives protection for the rest of the lifespan. Yet recovery from one of these infections gives no protection against any of the others.

This indicates that, besides the body's non-specific immunity, there must be another form of immunity which can 'remember' a previous infection and respond more effectively on subsequent encounters and, secondly, can distinguish very precisely between one infectious organism and another. These two features are characteristic of a sophisticated type of immune response known as *adaptive immunity*. It recognises, in a highly specific way, each non-self cell or molecule it encounters and *adapts* at this first meeting so that subsequent encounters provoke a much more effective immune response. The response to the first encounter takes two to three weeks to reach its peak and may be too weak to prevent symptoms from developing. The response to subsequent encounters peaks after about seven days and may be sufficient to eliminate the infection before it gains a hold, or at least to reduce the symptoms markedly.

□ On the basis of this information about the adaptive immune response, what would you expect an immunising injection to contain, and how do you suppose it brings about protection?

■ The injection, or vaccine, contains a preparation of the disease-causing organism (or one very like it) that has been rendered harmless by chemicals, heat treatment or freeze-drying. The harmless organisms circulate in the body and are recognised by the immune system, which then adapts so that any subsequent encounters are met with a vigorous, protective response.

Adaptive immunity rests on a particular type of white blood cell, the lymphocyte (Figure 9.5), a rather nondescript small cell that circulates in large numbers in both the bloodstream and in the lymphatic system. The extensive network of lymphatic capillaries is dotted with small clumps of lymphocytes packed into glands known as lymph nodes, as shown in Figure 9.6. During infections these glands may swell up and give pain, as occurs with the glands in the neck during a sore throat. Lymphocytes (together with all other types of blood cell) are formed in the bone marrow by the division of multipotent stem cells.

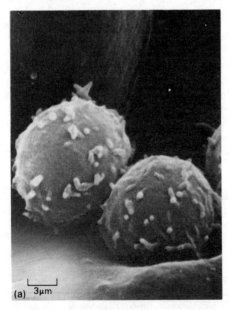

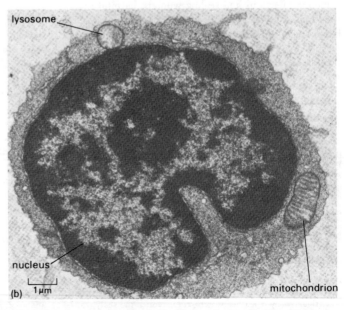

Figure 9.5 (a) Scanning electron microscope photograph of two lymphocytes, showing their ruffled surface membranes. (b) Electron microscope section through a lymphocyte. Notice that the nucleus fills most of the cell.

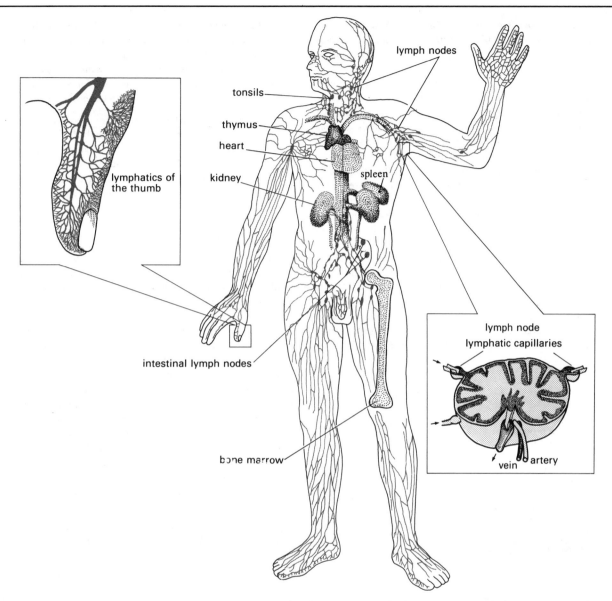

Figure 9.6 A simplified 'map' of the lymphatic system. All the cells in the lymphatic and blood system originate in the bone marrow, mainly in the long bones (shown here in red), the pelvis and breastbone. The actual extent of the capillary network can be visualised by looking at the inset drawing of the lymphatics of the thumb. The inset drawing on the right shows a schematic cross-section through a lymph node.

These stem cells contain more than 1 000 different genes involved in making proteins with various functions in the immune response. As daughter lymphocytes are generated by division of the parent cell, hundreds of these genes are removed from the DNA by special enzymes which then splice together the remaining genes. Each time a division takes place *different* sequences of genes are lost, or put another way, different sequences of genes *remain*. As a result of this 'sharing out' of available genes, each lymphocyte is programmed to recognise and respond to a *single* cluster of chemicals comprising no more than seven amino acids or simple sugars. The clusters to which lymphocytes respond are found on non-self cells and molecules, but not usually on self cell membranes or in the normal body fluids. Although the surfaces of bacteria, viruses, fungi and other microorganisms share many chemicals in common with those of human cells, scattered across the surface of these visitors are clusters that are unique to each type of organism and allow its precise recognition. Such a unique cluster is known as an *antigen* (see Figure 9.7) — it is equivalent to a fingerprint or name-tag.

Shuffling a relatively small number of stem cell genes and sharing them out in different combinations results in a huge diversification of the immune response. Although each individual lymphocyte can respond only to *one* antigen, the entire population of lymphocytes can, between them, respond to up to 1 000 million *different* antigens — more than enough to recognise all the pathogenic organisms on the planet. It is not known precisely how many lymphocytes start life sharing the *same* genes and hence recognising the *same* antigen, but there is not space in the circulation of a newborn baby to pack in more than a few of each.

Once humans leave the protected environment of the womb, they are exposed to antigens not only on microorganisms, but also in their food — in cow's milk, for example. When a lymphocyte encounters its matching antigen, it undergoes repeated cell divisions to form a *clone* of cells (a large population of genetically identical daughter cells) committed to attacking that single antigen. All other

molecules are ignored. This expansion takes two to three weeks, but after the first encounter has been overcome some of the expanded population survives to ensure a more rapid and effective response if the same antigen is encountered again. Any subsequent exposures leave the appropriate clone of lymphocytes even larger than before.

☐ On the basis of what you now know about adaptive immunity, why do you suppose that newborn babies are so vulnerable to infection and, for example, require feeding bottles to be carefully cleaned?

■ At birth, the baby's immune system is 'naive' since it has not encountered any of the potentially harmful organisms that inhabit its new environment. It can therefore mount only the relatively slow and weak first response to any infection.

Despite their uniform appearance, lymphocytes are divided into two main families, with very different functions. One family matures in the bone marrow and is known as B lymphocytes or simply B cells (B for *bone* marrow). They synthesise specialised proteins known as *antibodies*, which can bind very specifically to the antigens on non-self material. Note that each B lymphocyte can make only *one* kind of antibody that binds to *one* type of antigen. The other family is the T lymphocytes or T cells (T for *thymus*, a gland above the heart in which these cells mature). Some T cells can regulate the amount of antibodies produced by activating or suppressing the B cells, other T cells kill microorganisms directly. Thus T cells are divided into three main subsets, shown in Figure 9.8: (i) helper T cells (activate B cells to produce antibodies); (ii) suppressor T cells (inhibit B cells from producing antibodies); (iii) cytotoxic T cells (can kill directly without use of anti-

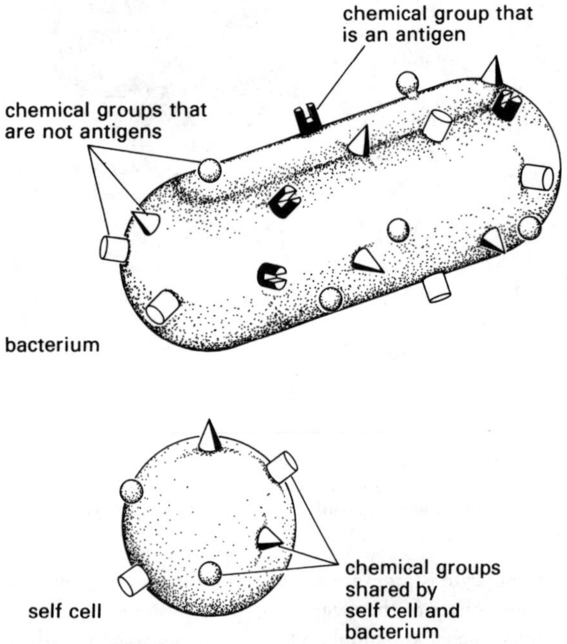

Figure 9.7 Some chemical groups on the surface of non-self cells (such as bacteria) *also* occur on human cells. These groups therefore do not provoke an immune response, since they are not perceived as antigens. An antigen is a chemical group that is normally found only on non-self cells. (The size of the chemical groups has been grossly exaggerated in this highly schematic illustration.)

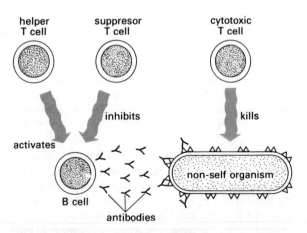

Figure 9.8 The interaction of three different types of T cells, and B cells, in the production of an adaptive immune response to a non-self organism. The helper and suppressor T cells regulate the amount of antibodies that B cells produce.

bodies). The proportions of each of these three T cell subsets is critical. There are about twice as many helper T cells as suppressor T cells in a healthy person, but in certain people the ratio may be severely disturbed, with disastrous effects on their ability to withstand infection. This is so in sufferers from acquired immune deficiency syndrome (AIDS), in which suppressor T cells far outnumber helper T cells, and immune responsiveness is therefore almost totally inhibited. Most AIDS sufferers die, within two years of diagnosis, from numerous unusual infections. This illustrates how dependent humans are on their immune system for protection.

Having recognised a pathogen, how does the adaptive immunity system achieve protection? As has already been mentioned, the cytotoxic T cells can kill microorganisms directly. This is achieved by secreting toxic substances on to their surface membranes. These T cells are particularly effective against viruses, even when the virus is sheltering inside self cells.

Antibodies produced by B cells are most successful against bacteria. They bind on to the surface of a bacterium via a 'lock and key' configuration in which part of the antibody molecule exactly fits the cluster of chemicals that form the antigen on the bacterial membrane (Figure 9.9).

The antibody itself causes no harm to the bacterium, but effectively labels it as a potential target for the phagocytic cells, the macrophages, to lock onto the bacterial membrane, and engulf it. In addition, the binding of antibodies to antigens triggers the activation of complement (the sequence of enzymes mentioned earlier) which results in damage to the bacterial membrane. To a lesser extent, viruses, fungi and pathogenic worms may all be damaged by these mechanisms, which depend on close interaction between the adaptive immunity and non-specific immune mechanisms and the inflammatory response, producing effective protection as a result of this collaboration.

The ability of an individual to make an immune response to a particular antigen does not only depend on whether or not they have encountered that antigen before. As with all other aspects of physiology, there are individual variations in so-called *immune responsiveness* which stem partly from the particular genetic make-up of that person. You probably know someone who always seems to catch every infection, or who has had, say, rubella several times. Even within the *same* person, immune responsiveness fluctuates according to other aspects of their physiological state, particularly the functioning of the nervous and hormonal systems. In recent years it has become clear that there are complex interactions between these three systems. For example, raised levels of the adrenal hormones, corticosteroids, have the effect of depressing the immune response to some extent, as do the sex hormones. Conversely, growth hormone, insulin and thyroxine increase immune responsiveness as their levels rise. Various neurotransmitters have also been shown to enhance or depress immunity and, in turn, molecules secreted by white blood cells involved in an immune response have an effect on activity in the nervous system. Thus the brain, immune system and hormonal system are linked in regulatory feedback circuits that 'modulate' immune responsiveness from day to day.

The immune system may become less effective in some individuals as they grow older, and may even 'make mistakes' and mount an immune response against self cells or molecules. A number of diseases, for example, rheumatoid arthritis, do have such an anti-self (or *auto-immune*) component and others, possibly including multiple sclerosis, may well include this type of immune response. In addition, an immune response against a genuine antigen may be excessively violent, and cause damage which in turn triggers off an inflammatory response (as in allergies). The severe tissue destruction seen in leprosy and tuberculosis is partly due to the presence in the infected area of toxic chemicals secreted by cytotoxic T lymphocytes.

The immune system has evolved 'hand in hand' with the evolution of pathogenic organisms, and these in turn have evolved various 'escape' mechanisms to evade destruction. For the moment, you should recognise that a dynamic balance exists between the ability of potentially harmful organisms to multiply and the immune system that seeks to control their numbers.

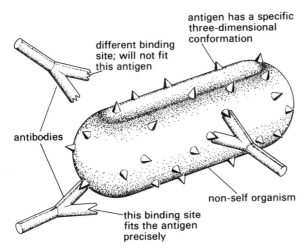

Figure 9.9 Schematic illustration of the 'lock and key' fit between a particular antibody molecule and its matching antigen. This enables the adaptive immune response to be highly specific for a given antigen.

Objectives for Chapter 9

When you have studied this chapter, you should be able to:

9.1 Describe the beneficial features of the inflammatory and the immune responses, and illustrate their potential for damaging healthy tissue if regulatory mechanisms fail.

9.2 Distinguish between the non-specific and adaptive immune responses, and illustrate their interdependence.

9.3 Use your knowledge of the adaptive immune response to deduce the biological basis of immunisation and the failure of so-called 'childhood infections' to recur in adults.

Questions for Chapter 9

1 (*Objective 9.1*) Three rare defects of the immune response (described below) serve to illustrate its normal capacity to protect us from infection. For each, describe which beneficial feature of normal immunity would be absent in an affected person and with what outcome.

(a) A disorder in which the phagocytic cells (for example, macrophages) can no longer produce hydrogen peroxide (bleach).

(b) A disorder in which B cells fail to mature and hence no antibodies are produced.

(c) A disorder in which the thymus gland never develops and hence no T cells can be formed.

2 (*Objective 9.2*) Choose one of the phrases (i)–(iii) to start each of the incomplete sentences (a)–(f). Which of the completed sentences illustrates the interdependence of the two response systems?

(i) Adaptive immunity

(ii) Non-specific immunity

(iii) Both adaptive and non-specific immunity

(a) ... can distinguish between self and non-self cells and molecules.

(b) ... cannot mount an enhanced response if exposed to a particular antigen for the second time.

(c) ... can distinguish between self and non-self material, but cannot distinguish between different types of bacteria or different virus particles.

(d) ... can retain a 'memory' of having encountered a particular antigen, in the sense that subsequent exposure produces an enhanced response.

(e) ... can produce specialised proteins (antibodies) that bind specifically with a particular antigen.

(f) ... can eliminate some non-self materials by engulfing them (phagocytosis), a process that is made easier if the antigen has antibodies already stuck to it.

3 (*Objective 9.3*) Immunisation against a particular infectious organism almost always takes the form of two or more injections or oral doses of vaccine at an interval of several weeks or months. Use your knowledge of the immune system to explain why more than one dose is given and why someone may have a single 'booster' dose in adult life shortly before travelling to an area that is heavily infected with that organism.

10
Reproduction

The fundamental needs of all species include the need to reproduce. In this chapter, we shall look at the physiological systems that have evolved in association with this need. We shall accept the existence of two distinct sexes as an evolutionary development without further discussion of the benefits that this brings. Some mention of the evolutionary advantages was made in Chapter 4 and these are discussed again in the next chapter. We turn now to the question of *how* the two sexes develop differently.

Human males and females differ anatomically in several respects, in addition to the basic difference of producing sperm or ova. How do these differences develop? Babies are usually clearly identified as girls or boys when they are born, so most of the biological development into two sexes must have occurred much earlier than birth, during fetal life. There are three factors that can influence the development of a human fetus into one sex or the other. These are the sex chromosomes, certain hormones and, to a lesser extent, the environment surrounding the fetus before birth. We shall consider each of these in turn.

Sex chromosomes
As explained in Chapter 4, ova can only have an X-chromosome, since females are usually XX. Sperm can have *either* an X-chromosome *or* a Y-chromosome. If an 'X sperm' fertilises the ovum, the fetus will be XX (female) and if a 'Y sperm' fertilises the ovum, the fetus will be XY (male). Although it is the sex chromosome of the fertilising sperm (X or Y) that determines the fetus' sex, *which* sperm actually fertilises the ovum from the many millions ejaculated may depend upon the environment. For example, the speed at which X or Y sperm swim through the female reproductive tract can be influenced by the degree of acidity in the woman's vagina.

As you have seen, the DNA in chromosomes directs the synthesis of proteins, including all enzymes, and hence controls chemical processes in the body. The processes the sex chromosomes control include the development of the gonads (ovaries or testes), the *primary sexual characteristics*. A few genes on other chromosomes also contribute.

Sex hormones
The term sex hormones is usually used to describe the hormones produced by the gonads. There are three main types: the androgens, produced mainly by the testes, and the oestrogens and progestogens, produced mainly by the ovaries. Both sexes produce all three kinds, even during fetal life, so it is not strictly accurate to call them 'male' and 'female' hormones. What differs between the sexes is the quantity of each kind produced. As early as the seventh to eighth week after conception, the fetus has either testes or ovaries, and the ratio of the hormones they produce will differ. The most important of these hormones for further sexual development are the androgens, produced mainly by the fetal testes. If there are relatively high levels, then male structures (such as the penis) develop; if there are very low levels (as there would be in females since they lack testes), then female structures (such as the vagina) develop. Oestrogens and progestogens have only a slight effect on sexual development in the fetus, but, as you will see shortly, they are important in the further sexual development of females after puberty.

☐ Can you think of a reason why oestrogens and progestogens are *not* responsible for sexual development in the fetus?

■ The fetus is likely to be exposed to quite high levels of oestrogens and progestogens, simply because it is developing inside an adult female body. Since *all* fetuses are thus exposed to these hormones, they would be of little use for developing sex *differences*. Differentiation of the two sexes therefore must depend upon hormones produced by the fetus itself.

Environment
Although the sex chromosomes and the sex hormones are the major factors which influence the sexual development of the fetus, the environment in which it grows can have an effect. Some hormonal drugs which were given to women during the 1940s (particularly in the USA) were sometimes found to provoke abnormalities in the sexual development of the fetus. However, these environmental effects on sexual development before birth are quite small compared with the role of the sex chromosomes and hormones. In contrast, other aspects of development can be severely affected by environmental 'insults' such as X-rays and drugs (for example, thalidomide).

After birth the baby is almost always recognisably female or male and, in nearly all human societies, will be brought up differently according to its recognised sex, eventually accepting herself or himself as belonging to one sex or the other. The term gender is sometimes used to describe the learnt, social aspects of becoming women or men as opposed to biological sex differences.

Puberty

The next phase of life at which females and males go through noticeably different stages, in biological terms, is at puberty, which usually starts somewhere between the ages of 10 and 15, and takes several years to complete. Puberty results in the ability to produce fertile ova or sperm, that is, reproductive maturity. During this time the development of *secondary sexual characteristics* occurs. In females, breast and genital development occur, and the menstrual cycle begins. In males, enlargement of the genitals and a deepening of the voice takes place. In both sexes, emergence of body hair, development of sweat glands, especially under the arms, and an increase in growth rate also occur.

Although nearly all individuals go through the bodily changes of puberty, they do not all go through them at the same ages. This variation can have profound social and emotional consequences.* All these changes are under the control of increasing levels of sex hormones secreted by the gonads. The levels of these hormones increase at a certain age because they are under the control of the pituitary gland. During childhood, the output of the pituitary hormones that stimulate the gonads (termed gonadotrophins) is low, but rises at a variable age, usually between 10 and 15 years, initiating the physical changes of puberty. This raises the question: why does the pituitary gland increase its stimulation of the gonads at this age? Only a partial answer can be given at present. It is known that a feedback loop exists (Figure 10.1) in which a rise in

* These are dealt with in *Birth to Old Age: Health in Transition*. The Open University (1985) *Birth to Old Age: Health in Transition*, The Open University Press (U205 *Health and Disease*, Book V).

the sex hormones in the bloodstream damps down the stimulatory hormones released by the pituitary gland. This enables very fine balancing of hormone levels in a stable state over a long period.

During childhood, the pituitary gland is very sensitive to the level of circulating sex hormones and even a tiny increase is sufficient to inhibit the release of the pituitary hormones that in turn stimulate the gonads. Thus the levels of both sets of hormones (pituitary and gonadal) are kept low. During puberty, the pituitary gland gradually loses its sensitivity to the sex hormones and 'allows' levels to rise much higher before its own output is inhibited. In fully reproductive adults, the pituitary gland is relatively insensitive to sex hormone levels. Why its sensitivity should change in this way is a mystery, but there are clues that point to 'higher' centres in the brain being involved.

The reproductive systems

The component organs of the male and female reproductive systems are shown in Figures 10.2 and 10.3. These organs are in close proximity to the ureters, which lead

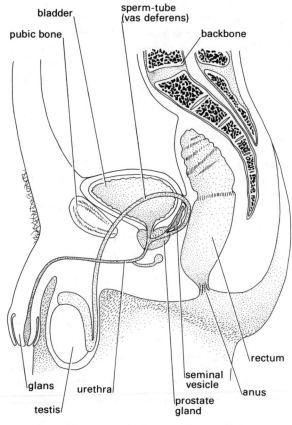

Figure 10.2 Section through the body showing one side of the male reproductive system (side view).

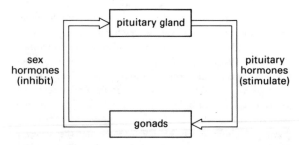

Figure 10.1 Feedback loop controlling the level of hormones released from the gonads.

from the kidneys to the bladder, and to the urethra. This has consequences for urinary tract infections, as you will see in Chapter 14.

☐ List the major functions that you think the male and female reproductive organs have to carry out.
■ (i) Produce ova or sperm; (ii) provide structures that allow fertilisation to take place inside the female body (the vagina and the penis); (iii) provide an environment for the development of the fertilised ovum into a baby (the uterus); (iv) provide a canal for the baby to emerge.

In the male, sperm are produced continuously throughout adult life inside the two testes (singular, testis). They then pass through a tube, the vas deferens, leading to the two seminal vesicles in which they are stored. Ejaculation is the process by which semen is propelled into the urethra by contraction of the muscles around the vas deferens and then expelled from the urethra by rhythmic contractions. Immediately prior to ejaculation, the sperms pass from the seminal vesicles into the urethra, joining it at a fork around which lies the prostate gland (Figure 10.2). This gland secretes a milky fluid which mixes with sperm and seminal fluid to help the sperm swim and provides nutrients.

Ova production is carried out in the two ovaries (Figure 10.3). Unlike sperm production, ova production ceases at

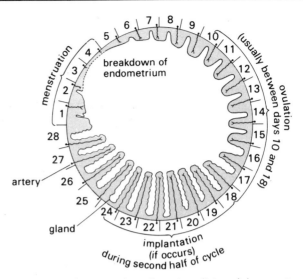

Figure 10.4 Changes in the endometrium (lining of the uterus) during the menstrual cycle in which day 1 marks the start of menstruation. Note that although the figure represents a cycle of 28 days, normal cycles may be as short as 21 days or as long as 35 days.

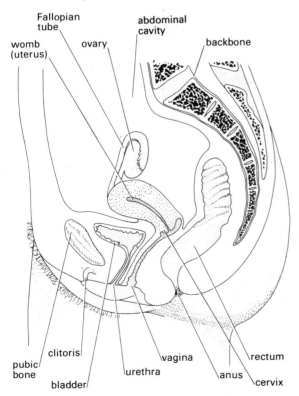

Figure 10.3 Female reproductive system (side view).

about the age of fifty. When a human female is born, she *already* has in her ovaries her entire stock of ova, but in an immature state. Throughout her fertile adult life, she will normally bring some of these immature ova to maturity each month. The ovum grows to maturity inside a protective and nutritive hollow ball of cells, which then moves towards the edge of the ovary, and ruptures. The mature ovum is then ejected from the ovary and is wafted by the movements of tiny hairs into the Fallopian tube which leads to the uterus. The uterus in turn leads to the vagina at the lower end, through the cervix, a muscular ring that protects the neck of the womb.

The regular monthly production of a fertile ovum is known as the *menstrual cycle*, and is under the control of pituitary and ovarian hormones. During the first half of the cycle, gonadotrophins from the pituitary gland bring a number of ova to maturity, but normally only one is released from the ovary. These hormones also stimulate the wall of the uterus to begin thickening and to become rich in blood vessels. The ovum is released from the ovary (ovulation) approximately half-way through the menstrual cycle and the ovary begins to secrete both oestrogens and progestogens, which accelerate these changes in the uterine wall. This build-up, shown in Figure 10.4, is necessary so that the uterine wall is both thick enough and sufficiently well supplied with blood vessels to afford nourishment to a fertilised ovum once it has implanted. If no ovum is implanted, the production of ovarian hormones declines, and the thickened walls of the uterus die, and are lost, together with blood, as the menstrual flow. This process is known as menstruation. If an ovum is implanted, the level of these hormones remains high during pregnancy and inhibits the production of further ova.

Since the late 1950s, the principles of hormonal control of the menstrual cycle have been used as the basis of the oral contraceptive pill. The most commonly used pill contains a combination of oestrogen and progestogen, which act mainly by preventing ovulation.

The onset of menstruation at puberty is termed menarche. Although menarche is taken as a sign that a girl is approaching biological maturity, it does not signify an immediate change to adult patterns of hormone secretion. Rather, pituitary hormones increase very gradually before the first menstrual period and continue to increase for some months thereafter. Her first few menstrual periods may not be very regular and it is usually some time afterwards that she becomes fully fertile. Indeed, less than half her menstrual cycles for the first year or so after menarche are likely to involve the actual release of an ovum. A similar period of relative infertility occurs during early puberty in boys — although in neither sex is such infertility predictable! Within a year or two of puberty, both sexes are fully fertile and capable of producing children.

Hormones and reproduction

Having looked at the basic layout of the reproductive systems, we must reconsider the sex hormones. As you have seen, the *balance* between the sex hormones, produced by the ovaries and testes, and the gonadotrophins, produced by the pituitary gland, is important during puberty, and remains crucial for the continued production of fertile ova and sperm throughout adult life. In addition, sexual responsiveness in men is roughly related to hormone levels. Castrated men lose their 'sex drive', but giving a man extra hormones does not necessarily increase it!

The hormonal balance is more complex in females because it is necessary to do two jobs: bring ova to maturity and help to prepare the uterus for any fertilised ovum. In women, hormone levels do not appear to affect sexuality: removal of the ovaries (which sometimes accompanies hysterectomy) does not normally affect women's 'sex drive'. However, the menstrual cycle can be disrupted during periods of emotional stress, by a change of sleeping pattern (for example, on journeys to a different time zone, or when beginning shift work), and by weight loss during periods of strict dieting. This is one of the clues that 'higher' centres in the brain may be involved in regulating hormone levels and, hence, the menstrual cycle.

Fertilisation and pregnancy

After sperm have been ejaculated into the vagina, they swim through a tiny opening in the cervix into the uterus.

From there, they pass into the Fallopian tubes, using their tails to propel them. It is normally in one of these tubes that the ovum is fertilised. As already discussed in Chapter 4, the nucleus of the fertilising sperm fuses with that of the ovum and the resulting cell has the full set (twenty-three pairs) of chromosomes. The fertilised ovum passes on down the Fallopian tube, dividing as it goes, and then implants into the wall of the uterus to continue development. However, sometimes a serious complication occurs when the ovum implants in the Fallopian tube (an ectopic pregnancy).

Implantation is the signal that pregnancy has begun and the mother's body becomes the environment for the developing fetus. The implanted ovum releases hormones which 'break' the normal rhythm of the menstrual cycle and cause the continued release of oestrogens and progestogens. Any further ovulation is therefore prevented from taking place.

The fetus is in constant interaction with its environment in the womb, and subject to various influences from it. What special structures are needed to protect it and supply its needs during growth? The primary protective structure is the amnion, a bag containing a watery fluid which keeps the fetus moist and protects it against shock or adhesions (Figure 10.5). Nourishment is provided partly by a yolk sac

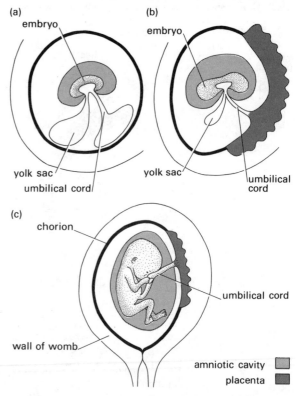

Figure 10.5 The development of an embryo and its life-support systems at approximately (a) four, (b) eight and (c) twelve weeks.

directly attached to the growing fetus. As its nutritive role is outlived, the sac is absorbed and incorporated into the gut. Mother and fetus are connected via the umbilical cord and the placenta, a structure which separates maternal and fetal blood supplies. There is no direct connection between maternal and embryonic circulations. Nutrients and oxygen pass from the mother's blood vessels through several intervening cell layers in the placenta to the blood of the fetus. Carbon dioxide and waste matter move in the opposite direction.

☐ Can you think why fetal and maternal blood and other systems are not directly linked?

■ Non-self material, such as bacteria in the mother's blood, cannot directly infect the fetus. There are exceptions, however, including the bacterium which causes syphilis and various viruses. Chemicals in the mother's body, such as alcohol and drugs, are also 'filtered' to some extent by the absence of direct contacts, although they can, and do, cross the placenta. Babies born to mothers addicted to narcotics show acute withdrawal symptoms. The most important reason is that, because the fetus is genetically distinct from the mother, her immune system could reject the fetus and, if her white blood cells could get across the placenta, treat it as non-self.

Birth

When the development of the fetus (as outlined in Chapter 4) has proceeded sufficiently for it to function outside the maternal environment, birth occurs. The first signs of the approaching birth is the descent of the fetus' head into the mother's pelvis (Figure 10.6). This may occur several weeks before birth. Labour itself is divided into three stages. The first stage is accompanied by the onset of rhythmic contractions which dilate the cervix to give a continuous clear passage for the baby to pass through. This stage may take between three and twelve hours. The second stage may take from thirty minutes to one hour and involves the delivery itself, which is brought about by more frequent contractions and a conscious effort to push on the part of the mother. If it has not already done so in the first stage, the amniotic sac often ruptures (called the 'waters breaking') and a rush of amniotic fluid usually emerges at the same time that the baby's head emerges from the vagina. The third stage of labour may last about ten minutes and comprises further contractions to expel the placenta and fetal membranes (collectively called the afterbirth). Some bleeding from the uterus also occurs.

What initiates labour is not fully understood, but it

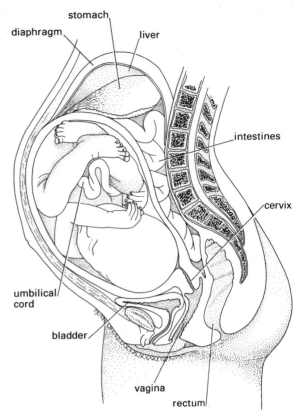

Figure 10.6 The mature fetus and parts of the mother's body shortly before birth (side view).

seems to involve a complex hormonal interplay between the mother and baby. It is thought that hormones from the baby's adrenal glands stimulate the mother's uterine wall to begin contractions. As labour progresses, however, more forceful contractions are needed to expel the baby. These are achieved by the production of another hormone, oxytocin, from the mother's pituitary gland. Synthetic versions of this hormone are used when labour is induced artificially.

Once the baby has been born, it is still in a very helpless state and continues to require warmth and nourishment. Human babies usually get the latter from maternal milk, produced under the stimulation of prolactin, another pituitary hormone. Suckling enables the continued production of oxytocin, which causes muscular contractions in the nipple to squirt milk into the baby's mouth. Oxytocin also promotes small contractions in the uterus which gradually returns it to its normal, non-pregnant, size.

Objectives for Chapter 10

When you have studied this chapter, you should be able to:

10.1 Describe the development of the reproductive system in humans, with reference to the role of chromosomes and hormones.

10.2 Describe the major biological features of the menstrual cycle, pregnancy and childbirth.

Questions for Chapter 10

1 (*Objective 10.1*) State whether (a)–(d) are true or false. Give reasons for your answers.

(a) Oestrogens are the hormones responsible for the development of primary sexual characteristics.

(b) The secondary sexual characteristics are a direct result of the action of pituitary hormones.

(c) New sperm and ova are continually produced throughout life.

(d) Both parents contribute to the sex of the offspring.

2 (*Objective 10.2*) For each item (a)–(e) select the most appropriate item from (i)–(v) to give a complete sentence.

(a) Prolactin ...

(b) A decline in oestrogen and progestogen levels ...

(c) Increase in gonadotrophins ...

(d) An increase in oestrogen and progestogen levels ...

(e) Oxytocin ...

(i) ... stimulates the ovary to produce its hormones.

(ii) ... stimulates uterine contractions.

(iii) ... causes menstrual bleeding.

(iv) ... helps to build up the womb lining.

(v) ... promotes the production of milk.

11
Introducing disease

In this chapter you will be referred to an article in the Course Reader by René Dubos, (Part 1, Section 1.1).

We now turn from considering the normal biological structure and function of the human body to the abnormal. Our concern must shift from physiology, or the normal functioning of cells, tissues and organs, to that of *pathology* — the disordered, abnormal structures and functions that underlie illness and disease. In the following five chapters many diseases will be described and discussed, but before we embark on this account of the biology of disease, we need to pause and consider three questions.

The first is, why does disease occur? You may already be familiar with the ways in which social scientists might respond to this question.* Here we consider some biological explanations. The other two questions arise as a result of the vast number of diseases that have been identified and described: how should we classify diseases so as to structure the discussion and which specific diseases should we include in an account that inevitably can mention only a very small proportion of the total number?

Why does disease occur?

Views of what causes disease have altered over the course of recorded history.† The most striking feature of these changes has been the shifting emphasis between the contribution of outside agents (such as evil spirits and germs) and that of defects within people. An example of the former was the rise of the germ theory in the nineteenth century, as a consequence of the work of microbiologists such as Pasteur and Koch. The fact that a microbe could strike down an apparently healthy person seemed to provide, at the time, irrefutable evidence that disease must

---

* The responses a social scientist might make to this question are dealt with in *Medical Knowledge: Doubt and Certainty* and *The Health of Nations*. The Open University (1985) *Medical Knowledge: Doubt and Certainty* and *The Health of Nations*, The Open University Press (U205 *Health and Disease*, Books II and III).

† Changes that occurred in the explanations of diabetes, plague and hysteria are described in two other books in the series *Studying Health and Disease* and *Medical Knowledge: Doubt and Certainty* (U205, *Health and Disease*, Books I and II).

come from the outside. In contrast, more recent work in the field of genetics has lent support to the view that susceptibility to disease is at least partly inherent in the individual. Although support for these two alternative views has shifted back and forth, at most times support for a contribution of both has existed. This was true even at the height of the germ theory at the end of the nineteenth century. This can be seen in descriptions of diseases at that time, such as the following account of tuberculosis.

> Consumption is decay of the lung substance, caused by an aggressive germ — the tubercle bacillus — dangerous to those wanting vital force. This germ has many inlets to attack the body through, and it gets to the blood perhaps by the air or food, but it goes no further than the blood. Pure blood contains white corpuscles, minute microscopic octopi, full of arms and legs, waiting to seize and devour the mischief-making germ. But if the patient's constitution is bad; if misery and poverty and overcrowding, or dissipating and enervating luxury, impair the powers of life, then the white corpuscles are lazy and languid and have no fight in them. The bacillus gets a foothold — its time has come, and it takes advantage of it. It selects the top bit of the lung — the apex as it is called — a portion stowed away in a corner, out of the way of the inspired air and the brisker circulation — a naturally weak spot, and the first to capitulate.
> (*The Chemist and Druggist*, 1898, p.38)

☐ In what ways does this description demonstrate a view of disease as being caused by external agents?

■ There are numerous phrases that suggest this: 'caused by an aggressive germ'; 'many inlets to attack the body'; 'white corpuscles ... waiting to seize and devour the mischief-making germ'.

☐ What evidence is there in the description which suggests that internal deficiencies were responsible?

■ Again, there are several phrases: 'if the patient's constitution is bad'; 'white corpuscles are lazy and languid'; 'a naturally weak spot'.

☐ In both Chapters 1 and 5 the use that biology makes of metaphors was discussed. What type of metaphors are used in this description?

■ Military ones: 'inlets to attack'; 'no fight in them'; 'the bacillus gets a foothold'; 'the first to capitulate'. The use of military metaphors was clearly well-suited to views of disease as the result of either an external 'attack' or a failure of internal 'defence'.

This focusing on external and internal factors has not, however, been the entire story of explanations for why disease occurs. Another long-standing belief has been that disease results not from outside forces alone, or simply from deficiencies in the affected individual, but rather as a consequence of the *interaction* of the two. This view was essentially that propounded by the ancient Chinese. For them, health reflected a proper balance between opposing forces called Yin and Yang, and disease was caused by a lack of balance between the two. The Ancient Greeks also saw health as a balance of reciprocal forces. Although this view has always attracted some support, during this century that support has grown considerably. In many ways the current interest in ecological theories of disease is the contemporary form of this tradition.*

> All living things, from man to the smallest microbe, live in association with other living things. Through the phenomena of biological evolution, an equilibrium is established which permits the different components of biochemical systems to live at peace together, indeed often to help one another. Whenever the equilibrium is disturbed by any means whatever either internal or external, one of the components of the system is favoured at the expense of the other ... and then come about the processes of disease. (Dubos, 1953, p.78)

☐ How do the metaphors Dubos uses compare with those of the ninteenth-century description of tuber-culosis?

■ Dubos uses metaphors of peace rather than war: 'live in association'; 'an equilibrium is established'; 'live at peace together'; 'often to help one another'.

Since the 1950s the earlier view of disease as being a consequence of lost battles has been giving way to the sorts of ecological view Dubos described. At the heart of such a view is the notion that an organism must maintain itself in a state of balance with the external environment if it is to survive. This idea of *external homeostasis* is seen as being just as important as the maintenance of *internal homeostasis*, described by Claude Bernard and other physiologists in the nineteenth century. Indeed, the two are intimately linked — failure of an organism to cope with changes in the external environment, both physical and social, will cause the internal mechanisms to act insufficiently or inappropriately, resulting in the produc-tion of disease. In other words, disease is seen as the result of the organism's failure to cope with the external physical and social environment. Dubos gives the example of tuberculosis, which, he concluded, resulted more from the 'social disruption' of the rapid social change that took place during industrialisation than from the insanitary con-ditions, crowding and poverty associated with that period.

* An example of this view is expressed by René Dubos in Part 1, Section 1.1 of the Course Reader. Black, Nick *et al.* (1984) *Health and Disease: A Reader*, The Open University Press.

More recently, studies of Jewish immigrants to Israel have shown the increased incidence of coronary heart disease in oriental Jews to be associated with the extraordinarily rapid social change that has occurred among them. Less dramatic, but equally supportive of this view, is the way that the *Herpes simplex* virus, harboured for years by healthy people, produces blisters (cold sores) at times of biological, social and emotional stress.

Although this ecological view of disease is the one to which many contemporary workers in this field subscribe, it is important that you realise that the military model persists. Indeed, there is still a dispute in biology and medicine as to which metaphor is the more appropriate. A similar dispute exists regarding explanations of human social behaviour: there are two main theories, functionalism and conflict theory.*

☐   What are the broad premises on which each of these theories are based?

■   Functionalism views an individual's interests as tending to generate a harmonious society, whereas conflict theory is based on the premise that society tends to contain groups with opposed and unreconciled interests.

Putting these theories in such stark terms runs the risk of oversimplification. Nevertheless, there are clear parallels between these social theories and the two contrasting biological views discussed above. This serves to underlie the similarities between the debates within both natural and social sciences.

Elsewhere in this course we have stressed the need to consider the historical context of current views, whether they be of the practice of medicine, beliefs about disease, or the social structure of different communities. The same is true of the biological world, for this too has changed over time. To understand the history of the biology of disease we must consider how disease, and the external homeostasis on which health depends, has evolved.

Evolution and disease

Evolutionary biology has two major concerns, phylogeny and adaptation. The study of phylogeny seeks to trace lines of descent from one species to another and from one group of organisms to another and tries to establish which extinct species and groups were the ancestors of contemporary species and groups. Here, however, we shall be concerned not with questions about distant ancestors, but with *adaptation*, the key concept in understanding how diseases may have evolved.

*These are discussed in *Studying Health and Disease* (U205, Book I).

When biologists say that an organism possesses a particular adaptation, they mean that it possesses a feature that makes it better able to fulfil some essential aspect of life, such as finding food, rearing its young or avoiding being eaten by predators. *Natural selection* acts on the differences between *individual* organisms: those that possess better adaptations are more likely to survive the rigours of life and are, therefore, more likely to breed and leave offspring in the next generation. If the differences upon which natural selection acts have some genetic basis, they will be passed on from parent to progeny. The process of natural selection is commonly summed up in the phrase 'the survival of the fittest', first used in the nineteenth century by the British philosopher and sociologist, Herbert Spencer. To Spencer, 'fit' meant strong, healthy and swift, as it does today in everyday speech. For evolutionary biologists, however, the word has come to have a rather different meaning and now refers to the number of offspring left by an individual. Charles Darwin, the nineteenth-century British naturalist whose evolutionary theory survives, with modifications, to the present day, repeatedly stressed that natural selection acts on *variation* in the ability, not simply to survive, but to survive *and* reproduce. Thus, a more accurate phrase to sum up the process of natural selection would be the 'reproduction of the fittest'.

If such a theory is correct, why is it that humans continue to suffer from diseases? Surely they should have evolved a perfect form of adaptation so that they can live in complete harmony with the environment, free from all illness. There seem to be two answers to this question. First, that it is not within the power of natural selection to prevent illness and death. Second, it is necessary to consider not only the evolution of humans, but that of other species as well. After all, what is a disease to a human is a way of life for the microorganism involved!

Before considering these two explanations in greater detail, we must mention two others which have been proposed. The first is that the task of producing perfectly adapted humans by natural selection is, as yet, incomplete and that this will be achieved at some point in the future. Evolutionary biologists do not subscribe to this view because they recognise that the environment in which organisms, including humans, live is constantly changing. Consequently, over the generations, slightly different adaptations are favoured so that natural selection, in a sense, will inevitably always lag slightly behind environmental changes. The second is that disease and death must be adaptive characteristics. This may at first seem absurd. However, it has been suggested that it is adaptive for adult animals and plants to die after they have reproduced because, by so doing, they do not compete with their progeny for limited resources and the progeny are therefore more likely to survive. This may be true for some species, but to many biologists, this suggestion — that illness and death are adaptive — is an example of a form of fallacious reasoning called adaptationism. They reject the assumption that, *because* natural selection produces adaptation, any character that an organism possesses *must* be adaptive.

We now need to consider the two explanations that are generally regarded as explaining why humans have not evolved into a disease-free, immortal state. The first of these is that such an expectation is based on false assumptions about the power of natural selection. This can be demonstrated by considering the effect of natural selection on the *gene pool* of a population: that is, the sum total of the genes possessed by all the individuals making up the population. To be more specific, we must consider those genes that cause or hasten death, and consider them in two categories based on whether they are expressed (i.e. actively concerned in synthesising protein) before or after someone reproduces. Let us consider, first, potentially lethal genes that are expressed *before* a person reproduces. These cannot be passed on to their children (because the person dies young) and so there is very powerful selection operating against them, which would lead to their elimination from the gene pool in the population. This has traditionally been true of fatal diseases with some genetic basis which start in childhood, such as cystic fibrosis and Type I (insulin-dependent) diabetes. Children suffering from these conditions frequently died before reaching adulthood. In other words, it would appear that natural selection has helped to prevent these genes being passed on to future generations, thereby minimising the incidence of fatal diseases and deaths in childhood.

In passing, it is of interest to speculate as to the effect that the present-day survival of such children (as a result of medical treatment) will have on the gene pool of the population. If genes are only one determinant of Type I diabetes, it seems unlikely that the incidence of the condition in future generations would alter much. However, recent reports that the incidence of diabetes appears to have risen over the last few decades has raised the question as to whether an increase in the frequency of the relevant genes could have been responsible.

Let us now consider the other case of potentially lethal genes, those that are expressed *after* a person has reproduced. Clearly, all genes in this category will be passed on to the children and will not be eliminated from the gene pool. In other words, some genes that cause death will be retained in the gene pool. We can demonstrate this process by considering the effect that natural selection has on individuals at different ages and, therefore, with different numbers of children. Consider someone aged 30 who has one child. Suppose that sometime during the next five years this person has one more child: he or she would have doubled the number of their genes passed on to the next generation. Now suppose that, instead of surviving those five years and having a second child, the person had died at the age of 30, as a result of natural selection operating against unfavourable genes.

☐ What impact would that person's early death at 30 have had on the gene pool?
■ It would have halved the number of genes he or she passed on to the next generation.

Now let us suppose that same person lived on to 40, by which age he or she had three children.

☐ What is the effect on the gene pool of this person dying at the age of 40, compared with living on and having another child?
■ Not having a fourth child would reduce the number of genes passed on by a quarter (one child out of four).

Clearly, the age at which someone dies influences their *absolute* contribution to the gene pool: the younger they are when they die, the fewer will be the number of children that survive them. What about the *relative*, rather than absolute, difference that age of death makes to this process?

☐ In other words, does the effect of natural selection on the frequency of particular genes in the gene pool become more or less powerful with increasing age of death? (Using the example, you need to compare the relative effects of dying at 30 and at 40 years of age.)
■ The effect becomes less powerful with increasing age. In the example, death at 30 halved the number of genes passed on whereas death at 40 reduced it by only a quarter.

If the effect of natural selection in eliminating unfavourable genes decreases as people grow older, then (in theory) a point would be reached where natural selection would have no effect at all. On this kind of argument, then, ageing and death are *an inevitable consequence* of the fact that natural selection *cannot* prevent them. In other words, it is not surprising that humans have not attained a disease-free and immortal existence as a result of natural selection. Although natural selection has the power to minimise fatal diseases in childhood, its inability to affect the incidence of disease in adults means that it is unable to extend the lifespan.

Co-evolution

We must now consider the second explanation offered as to why humans have not attained a disease-free existence: that based on the fact that humans do not live and evolve independently of other organisms. Part of the ecological view of health is that survival is dependent on the preservation and maintenance of harmony, or external homeostasis. Clearly, the same criteria apply to the very microorganisms that may cause disease in humans. Thus, if we are to understand why infectious diseases occur, we must consider not only human evolution, but also that of the infecting organisms.

If a species of antelope evolves a running speed such that it becomes less likely to be caught by predators, those predators will also tend to evolve greater speed or greater cunning so that they can continue to find sufficient food. The study of such interactions is called *co-evolution*. Although natural selection will favour adaptations of host organisms that protect them from microorganisms, it also favours adaptations of those microorganisms that may in turn enable them to harm their hosts. The process of co-evolution is sometimes referred to as an evolutionary 'arms race', though this is unfortunate because the human arms race has properties that have no counterpart in the natural world.

In the co-evolution of microorganisms and humans, the former typically have one huge advantage: they have a very much shorter generation time than their human host. As you have seen, bacteria produce many generations in the course of an hour, whereas the human generation time is about twenty years. This means that most microorganisms can evolve much faster, in real time, than humans. We might expect, therefore, that human evolution would never be able to keep up. Why is it, then, that humans survive despite being subject to infection from a wide range of microorganisms?

There are good reasons for supposing that, in fact, microorganisms will not evolve increasing *virulence* (the ability to cause disease). Individual pathogens that kill their host quickly will tend to leave fewer offspring than those whose hosts live for a long time. Furthermore, it may be difficult for them to find a new host when their present one has died, either because hosts are widely dispersed or because intermediate hosts* are rare. Hence microorganisms are at risk of dying out themselves if they kill their host. There is thus a general argument in the theory of co-evolution that selection will favour not only increased resistance in hosts, but also decreased virulence in microorganisms. (If this is true, it illustrates the inappropriateness of the arms race analogy.) The notion that such a 'balance' is struck between host and microorganisms is well illustrated by the devastation a microorganism can cause if it becomes able to infect a species *other* than its usual host, as in the case of plague,

which is far more lethal to humans than to rats. Other examples are rabies in foxes, yellow fever in monkeys, and brucellosis in cattle. All seem to have established a balance during thousands of years of co-evolution, but are often fatal to their more recent human host.

Having said that, it must be remembered that one of the most striking features of evolution is the enormous *diversity* that it has produced: among organisms living in apparently similar environments, we find many different forms of adaptation. For example, insects, bats and birds have all evolved flight, but possess wings that are constructed in totally different ways. A trend towards reduced virulence may be true for many microorganisms, but it is not the only adaptive pathway they might take. For example, if microorganisms are able to survive for a long time outside a host, then it will not be particularly costly to them if they kill their host, and there would then be little advantage in evolving reduced virulence. Some bacteria do seem to have evolved in this way and to have remained extremely virulent: for example, anthrax can survive in soil for many years. This is well illustrated by the example of Gruinard, a small island off the Scottish coast that was used for military experiments in the development of anthrax as a biological weapon during the Second World War. Even today, the island remains sealed off because of the continuing risk of contracting anthrax.

The theory of co-evolution, therefore, suggests that microorganisms and humans will evolve in relation to changes in their wider environment. There is abundant evidence from both animals and humans that, over the course of time, the impact that a microorganism has on its host changes markedly as a result of natural selection. This may explain some of the unresolved mysteries of changes in the patterns of communicable disease in humans, such as the demise of plague in the seventeenth century and of scarlet fever in the nineteenth century.

One of the best documented examples of co-evolution is that of the gene, or trait, for sickle-cell anaemia in people of African descent. As explained in Chapter 4, the sickle-cell gene is recessive. This means that the disease of sickle-cell anaemia occurs only in homozygous people, those who have two copies of the sickle haemoglobin gene. Although the disease (sickle-cell anaemia) is relatively uncommon (about 1 in 400 people), the proportion of carriers (people with sickle-cell trait) is as high as 10 per cent of the population. These people are heterozygous: that is, they have one gene for normal haemoglobin and one gene for sickle haemoglobin. This proportion of the population appears to be extremely common for a gene that, in the offspring of two carriers, may prove lethal. The explanation is that carriers benefit from an enhanced protection against malaria. In other words, possession of the trait confers some survival advantage to 1 in 10 of the

* Intermediate hosts refers to other species which the microorganism also infects.

population which more than compensates for the dangers faced by the 1 in 400 with sickle-cell anaemia. This situation has arisen as a result of the co-evolution of humans and the protozoan that causes malaria, *Plasmodium.*

The co-evolution of microorganism and host is often most clearly revealed when a microorganism is taken to a part of the world where it did not previously exist and where, consequently, the host has had no chance to adapt. Consider, for instance, the decisive impact of Old World diseases on New World communities during their first encounter in the fifteenth and sixteenth centuries.

Other species besides humans have also experienced such disasters. The classic example is that of the *Myxoma* virus and the rabbit. This virus is indigenous to South America and rabbits there are prone to the disease that it causes — myxomatosis, a form of fibrous skin cancer — though they do not usually die from it. The virus was unknown in other parts of the world until it was deliberately introduced into Australia in 1950, in the State of Victoria, in an attempt to control the rabbit population, which was threatening the sheep farmers' livelihood. Rabbits had themselves been introduced to Australia from Europe in the nineteenth century and had undergone a population explosion. The effect on the unprepared Australian rabbit was devastating. Carried by mosquitos, the *Myxoma* virus caused an epidemic that killed 99.8 per cent of all the rabbits and a second epidemic killed 90 per cent of the surviving population. However, the third epidemic led to only 50 per cent mortality in the remaining rabbits. Clearly, the disease was rapidly losing its efficacy as a means of controlling the rabbit population. This was due to the effect of natural selection on both the rabbits and the virus. Rabbits with inherited immunity to the disease, though initially very rare, survived and bred and, with little competition from other rabbits, spread very rapidly. In addition, those rabbits that contracted the disease, but did not die of it, were effectively immunised against it and survived subsequent outbreaks. At the same time, natural selection favoured less virulent strains of the virus, which consequently became less effective as a means of controlling rabbits. Today, rabbits have once again become a serious threat to Australian agriculture and new methods of control are being sought.

There is one other factor to consider in the context of co-evolution, that of the means of reproduction of different species. As mentioned earlier, the generation time of microorganisms is very much shorter than that of humans. So much so that, during the lifetime of humans, microorganisms that live within them will have completed many thousands of generations. As a result, microorganisms have an opportunity to evolve by natural selection during the lifetime of a single human, who cannot 'evolve'

but relies on the immediate responses made by the various biological defence mechanisms. By the time humans reproduce, their internal microorganisms have become better adapted to survive, but through sexual rather than asexual reproduction, humans produce genetically diverse progeny.

☐ How could sexual reproduction counteract the apparent advantage gained by microorganisms evolving during a human lifetime?

■ The progeny of sexual reproduction are genetically different from their parents and so have subtle differences in physiology and biochemistry. Although the microorganisms evolve adaptations that suit them to infecting the parents, they will be less well suited to infecting the progeny.

Thus, any microorganisms that are passed from parent to progeny will tend to be ill-adapted to their new hosts, who will therefore be less adversely affected by them. One of the adaptive features of sexual reproduction therefore may be a role in defending a species in its co-evolutionary relationship with microorganisms.

The classification of diseases

The classification of diseases has been a subject of discussion and debate ever since illness was recognised and defined in terms of discrete diseases. One of the first approaches to this task is associated with Thomas Sydenham in the seventeenth century. Belief in a natural order of diseases, similar to the patterns and hierarchies which were being imposed on plant species by botanists at that time, led physicians to attempt to do the same with diseases.

☐ Why did this prove to be impossible?

■ Disease categories (the equivalent of plant species) change as the criteria for defining disease alters. The categorisation of diseases can be seen to reflect the concerns of clinical medicine and medical research at any particular time.

The simplest system for classifying disease largely avoids the problems of shifting definitions by being based on the body system or organ in which the disease occurs. Most medical textbooks have adopted this approach. Diseases affecting the digestive system are grouped together in one section, which is then subdivided into diseases of the stomach, the intestines, the liver, etc. Diseases affecting a particular organ, such as the liver, may then be classified according to pathology.

However, when considering the pathology of a disease, we have to take into account the fact that diseases alter over time, that they commence, develop and pass through different stages. Some undergo periods of remission, during which the disease abates and health is restored for a while.

Thus to talk about the pathology of a disease as if it were a fixed state of affairs is too limited. Two further concepts are required. The process of the development of pathology is referred to as *pathogenesis*. Thus the pathogenesis of plague refers to the processes that occur in the cells, tissues and organs of the body between the arrival of the causative bacteria, *Yersinia pestis*, and the establishment of the pathological changes associated with the disease, such as the buboes.

The second aspect of the changing nature of pathology is the *natural history* of a disease. What doctors mean by that is the characteristic way in which a disease 'runs its course'. You will already be familiar with the natural history of many diseases.

☐ Describe the natural history of influenza.

■ It is a disease that comes on quite rapidly (between a few hours and a day) and lasts for several days during which generalised muscular aches, fever, headache and sore throat are suffered before a gradual restoration of health occurs. It may then take several more days of convalescence before health is fully recovered.

Classifications of diseases based on pathology usually recognise several main categories including *congenital defects* (those already present at the time of birth), excessive inflammation, abnormal cell growth and degeneration. A list of some of the diseases affecting the liver classified in this way is shown in Table 11.1.

Table 11.1 Diseases of the liver, classified on the basis of pathology

Pathology	Examples of diseases
congenital defects	congential cystic liver disease Dubin–Johnson disease Gilbert's disease
inflammation	viral hepatitis acute cholecystitis amoebic abscess
abnormal cell growth	hepatoma
degeneration	cirrhosis

This approach to classification reflects the traditional interests of medicine — the identification of diseased individuals, the diagnosis of their disease and intervention to cure or alleviate the symptoms. In other words, it is a classification suited to the practice of curative rather than preventive medicine. In contrast, the latter would require a classification that identified the biological and social *causes* of diseases as opposed to their pathological effects on the normal functioning of the body.

☐ Can you suggest some main categories of biological cause?

■ You may have thought of the following: genetic; infection; physical trauma; poisons; radiation.

Although the concern of this course is not restricted to the prevention of disease, we have nevertheless adopted a classification based on cause. This is for two reasons. The first is that one of the main aims of the course is to enable you to understand the factors that determine the occurrence of disease, and the second is that a cause-based classification is able to accommodate social determinants in addition to biological ones. In other words, such a classification allows us to pursue another aim: that of uniting material and knowledge from biological and social perspectives. Each discipline may offer a different explanation as to the cause of a particular disease.

☐ What is the cause of tuberculosis?

■ There are several possible answers, including the following: presence of the tubercle bacillus; low resistance to infection; poor nutrition; low income; inequitable distribution of resources both within society and between nations; periods of rapid social change.

It would appear that classifying diseases on the basis of cause is going to be difficult, or even impossible, unless we can first define what we mean by cause. In the following chapters we shall classify diseases in terms of their biological or *proximate cause*. In any chain of causation, there is one factor that occurs last, that is nearest to the onset of the disease. The proximate cause is the last step in the chain of causative factors that directly or immediately leads to the onset of pathogenesis.

☐ What is the proximate cause of tuberculosis?
■ Tubercle bacilli.

Distal causes are those that make up the chain of factors leading up to the proximate cause. These would include such factors as housing and nutrition in the case of tuberculosis.

☐ What are the commonest proximate and distal causes of lung cancer?
■ The commonest proximate cause is a chemical that induces cancerous changes in cells (a *carcinogen*) in tobacco smoke; the distal causes are the various social and personal reasons for people smoking cigarettes and genetic factors that might affect susceptibility.

Proximate and distal causes can be viewed as a chain or series of steps in the causation of disease, with the proximate cause being the final link or step before pathological changes occur in the body. That is not to say that the proximate cause is more important than the distal ones. If anything, the reverse is true. Consider again the example of tuberculosis. Tubercle bacilli may not harm a well-nourished, healthy person, whereas a distal cause of disease, such as poor nutrition, will almost certainly lead to disease of some sort, whether it be tuberculosis or some other infection. In other words, distal causes make humans susceptible to disease.

It is also important to note that distal causes are features of the breakdown of external and internal homeostasis. Absence of social harmony results in poverty and hardship; inadequate food intake results in damage to biological defence mechanisms, such as the inflammatory response. Indeed, the failure to identify the proximate cause of any of the conditions which are currently responsible for the majority of deaths in the UK has stimulated interest in the ecological approach to understanding disease. Conditions with unknown proximate cause have been grouped together and are discussed in Chapter 16. First, though, we shall discuss four other groups of diseases for which the proximate cause is known: diseases with a genetic or chromosomal basis (Chapter 12); diseases that are caused by physical agents in the environment such as radiation, extremes of temperature and physical trauma (Chapter 13); infections and infestations (Chapter 14); and diseases in which aspects of the social and psychological environment appear to be the proximate cause (Chapter 15).

The final category exemplifies the difficulties mentioned earlier of distinguishing between biological and social explanations. The interrelationship between the two has long been recognised. Indeed, medieval theory did not make such a distinction. In Elizabethan times the part played by emotions in causing illness can be seen in the writings of Thomas Wright.

> Passions cause many maladies, and well-nigh all are increased by them, for all that pain engendereth melancholy, which for the most part, nourisheth all diseases: for many we reade of that were cured by mirth, but never any by sorrow or heavinesse.
> (Wright, 1971, p.63, first published in 1604)

Thus, as you read through the following five chapters, it is important to remember that the classification we have adopted simplifies ideas of causation — a necessary device in order to avoid presenting an unstructured mass of diseases or attempting the presentation of multidimensional models of causation.

Finally, a brief explanation of how we selected the diseases that are discussed. There were two main criteria. Either the disease appears elsewhere in the course and it was felt that some knowledge of its biology was necessary, or the disease is a particularly vivid example of some biological process that we wish to explain. As you read on, you may notice that one important group of conditions is largely absent, that of mental illness. This is partly because the biological explanations for mental disorders are little

understood and partly because these disorders are discussed in some detail in *Experiencing and Explaining Disease.**

A few warnings

As was stressed in Chapter 1, biology is an experimental science — the facts come out of experiments and the theories have the same 'provisional' status as all scientific theories. This is as true for biomedical explanations of disease as for the natural sciences of genetics, cell biology, physiology and the social sciences. The descriptions of diseases and their causes which follow are inevitably based on the understanding of Western medicine in the early 1980s. Descriptions and explanations will change in the future, just as they have in the past. Such events give rise to the phenomenon of diseases that no longer appear to exist, such as the English Sweat.

Both theory and observation in biology can give rise to mistaken explanations. An example of the former is the influence that theories of hearing had on descriptions of the eardrum in the eighteenth century. Before the study of acoustics revealed how a diaphragm, such as the eardrum, could transform and transmit sound by mechanical means, it was generally recognised that there must, of necessity, be a hole present in the eardrum to maintain hearing. As a result, physicians all reported the presence of such a hole in people with normal hearing, despite the fact that no such hole exists. Once physicists had explained how a diaphragm was able to transmit sound, physicians stopped observing the 'hole'!

The influence of observation on theory can be equally misleading. In earlier chapters you read descriptions of intracellular structures. Much of that knowledge is derived from high-powered magnification of cells obtained using electron microscopes. To obtain such micrographs it is necessary to kill the cells with chemicals and rapid freezing, and then cut the material into extremely thin slices. In other words, the cells which are finally examined and described have been both chemically and physically altered. It is assumed that the appearance of cells in electron micrographs is a reliable representation of their natural living structure. If this assumption is wrong, then the theories of intracellular structure and function might be as distorted as the cells on which the theories are based. As with the warnings about the interpretation of historical, statistical and social data, it is important that you are aware that biology is equally susceptible to observational artefacts and theoretical errors.

---

* The Open University (1985) *Experiencing and Explaining Disease*, The Open University Press (U205 *Health and Disease*, Book VI).

Finally, a health warning. Descriptions of disease have the well-known effect of producing identical symptoms in the reader. The effect is greatest when the disease is particularly revolting, lethal or insidious in its onset. There is no known way of avoiding this phenomenon, but you might like to make a mental note that you felt quite healthy before you started on these chapters!

Objectives for Chapter 11

When you have studied this chapter, you should be able to:

11.1 Describe the two main biological views of why diseases occur, the 'military' and the ecological views.

11.2 Explain why humans will not evolve into a disease-free state as a result of natural selection, and why any consideration of evolution and infectious disease must take into account the evolution of organisms other than humans.

11.3 Explain what is meant by the pathology, pathogenesis and natural history of a disease.

11.4 Distinguish between proximate and distal causes of disease.

Questions for Chapter 11

1 (*Objective 11.1*) What is meant by the ecological view of disease?

2 (*Objective 11.2*) What are the two main reasons for believing that humans will not evolve into a disease-free, immortal state through the process of natural selection?

3 (*Objectives 11.3 and 11.4*) Acute bronchitis is an infection of the lungs. It has a sudden onset with cough, fever, and sometimes breathlessness and chest pain. It is caused by either a virus or a bacterium and occurs more commonly in people who smoke, suffer poor social conditions and come from families in which other members also suffer from bronchitis. Each attack of bronchitis causes further damage to the bronchi in the lungs. Identify the following features of acute bronchitis: (a) proximate cause(s); (b) distal cause(s); (c) permanent pathology; and (d) any aspects of the natural history.

12

Genetic factors in disease

This chapter builds on the discussion of the genetics of inheritance in Chapter 4. Various examples of inherited conditions are included for illustrative purposes and you are only expected to learn about the underlying mechanisms, not about specific diseases.

In this chapter, we are going to look at diseases which arise principally from a defect *within* the body, from an abnormality in a person's genes.

☐ Can you remember what a gene is?
■ A gene is a portion of a DNA molecule which codes for a protein. Occasionally, the protein coded for by one particular gene governs one character of an individual (eye colour, for instance). More often the products of many different genes interact with one another and with environmental factors to define one character.

Genetic make-up is one of the many factors that determine health. Variations in genes underlie some of the differences in appearance and metabolism that exist between people. Occasionally, a genetically determined character is so unusual as to be labelled abnormal. Albinos, for instance, have a gene that means they have no pigment at all in their skin. Some genetic abnormalities lead to disabilities; this is true for albinos, who suffer from sunburn with even mild exposure to sunlight. Such abnormalities, that can be passed on from generation to generation, are known as *inherited diseases*.

There are two main types: those that result from a defect in a single gene and those that result from defects in whole chromosomes. We shall start by considering single-gene defects.

Single-gene defects
Single-gene abnormalities arise as a result of mutations, changes in the structure of DNA during cell replication. If even one base of an active gene is changed, then one amino acid, in one protein produced by that cell, may be different. Such mistakes in DNA replication can occur spontaneously, but may also be induced by certain *mutagens*, such as radiation and some chemicals. If such a mutation occurs in only *one* cell in the body there may be no obvious disorder, though such genetic changes have been implicated in the initiation of cancers and this will be discussed more fully in

Chapter 16. If mutations occur in the formation of sperm or ova, then the results will affect many cells since a fertilised ovum will contain the altered gene, as will *all* the cells of the developing fetus and, ultimately, the adult. Whether or not a single abnormal gene results in disease depends on several factors.

☐ What do you think these might be?

■ 1 Whether the abnormal gene is dominant or recessive.

2 Whether or not the change in the protein (for which the altered gene carries the code) impairs normal function. Not all changes are harmful; some may be advantageous, many will be unnoticeable in their effect.

3 How widespread the effects of the impaired function of the altered protein are. If the defective protein is non-essential, then its malfunction may be unimportant. If it takes part in essential reactions in many cells, then its malfunction will have widespread effects in the body. The importance of any particular protein may depend on the environment: for instance, albinos suffer much less in temperate than in tropical climates.

Most single-gene alterations are either in genes that carry the code for proteins that are so important that the alteration is lethal early in development (resulting in the death of the fetus and spontaneous abortion), or else have little effect. However, a few result in well-recognised diseases. These diseases may, in principle, be caused by the presence of either one dominant or two recessive defective genes, but, in practice, only a minority of known single-gene defects are caused by dominant genes.

☐ Why do you think this might be? You need to consider the evolutionary aspects of disease discussed in Chapter 11.

■ A person with the dominant abnormal gene always has the disorder. If it is serious, they will be less likely to have children, either by choice or because they are too ill to reproduce (they may even die before reaching reproductive age). This means that the deleterious gene is not passed on to the next generation, but disappears when the affected person dies.

Because of this, most of the serious disorders caused by dominant genes are not passed on from generation to generation, but arise spontaneously and, therefore, their incidence depends on the mutation rate of the gene. This explains why the more commonly seen diseases caused by dominant-gene defects are less serious conditions or occur later in life. A consequence of this is that the sufferer's capacity to reproduce is not impaired and the disease may be passed on to children. In osteogenesis imperfecta (imperfect bone formation) the bones are abnormally fragile because of a defect in the formation of collagen, the protein that forms the connective tissue in bones and elsewhere in the body. The whites of the eyes of sufferers may appear bluish because the defective collagen in them reflects light differently from normal collagen. People with osteogenesis imperfecta may be severely disabled by the disease. One Viking chief in the ninth century, known as Ivar the Boneless, suffered from it and had to be carried into battle on a shield.

Another example of a dominant-gene defect that is passed down through families is Huntington's disease. This was first described in 1872 by a New England doctor, George Huntington, whose father and grandfather had also been doctors in the area and had observed the disease affecting three generations of one family. George Huntington described the gradual loss of intellect (dementia) that developed in middle-age, together with a defect in the coordination of movement which is now known to be due to a progressive degeneration in the nervous system. The underlying protein defect resulting from the genetic abnormality is not known. Someone who has Huntington's disease has one normal gene (*h*) and one abnormal gene (*H*) that is dominant.

☐ In the case of a man with Huntington's disease, which gene will his sperm contain?

■ Half his sperm will contain the normal gene (*h*) and half the abnormal gene (*H*).

☐ If this man with Huntington's disease has children with a woman who has two normal genes (*hh*), what chance is there that their child will have the disease?

■ It is easiest to write it like this:

Parents:	*Hh*		*hh*	
	affected		unaffected	
Sperm and ova:	*H*	*h*	*h*	*h*
Possible children:	*Hh*	*Hh*	*hh*	*hh*
	affected		unaffected	

You can see that half the children will have the defective gene, and therefore the disease.

Dominant-gene defects are usually easy to trace back through family trees, since the disease appears in every generation. By 1932, there were about 1 000 people in America known to suffer from Huntington's disease, and it could be shown that they were all descended directly from two brothers who came from Suffolk to Boston in 1630. Currently, the disease occurs in about 1 in 20 000 people.

We shall move on now to consider inherited disorders arising from recessive genes. Abnormal recessive genes cause disease only if the person is homozygous for that gene. The commonest disease, in people of European ancestry, caused by a single gene is cystic fibrosis (about 1 in 2 000 live births) in which the mucus produced in the lungs and pancreas is excessively thick and viscous. This frequently results in the affected child suffering from digestive disorders. Figure 12.1 shows the family tree, known as a pedigree, of a family where some members suffered from cystic fibrosis.

☐ What can you say about the genes of each of the healthy parents of an affected child?

■ They must each carry one gene for cystic fibrosis; that is, they are heterozygous. If you are uncertain about this, remember that each affected child has two cystic fibrosis genes, one on each of a pair of chromosomes. Each parent must have contributed one of these abnormal genes, so each parent must carry one gene for cystic fibrosis. You know that they do not have two cystic fibrosis genes, since they do not have the disease themselves.

This can be shown in the following way. If we represent the cystic fibrosis gene by f and a normal gene by F:

Parents:	Ff		Ff	
Sperm and ova:	F	f	F	f
Children:	FF	Ff	fF	ff

So, if two carriers of the cystic fibrosis gene have children together, then 1 in 4 of their children (on average) will have cystic fibrosis.

☐ On average, how many of their children will be carriers (heterozygous) for the gene?

■ 2 in 4, or half of their children.

Another example of a recessive disorder is Tay–Sachs disease, which has a particularly high incidence (1 in 2 000)

among Ashkenazi Jews whose origins are Eastern European. A defect in an enzyme that catalyses one step in the breakdown of fats leads to the accumulation of fat inside the cells in the brain and in the retina (at the back of the eye). Those cells become swollen and distorted. At as early as three months of age, signs of brain damage begin to appear — the child shows weakness in movement, followed by blindness and convulsions. Death usually occurs in early childhood. People who are heterozygous for the gene show no sign of disease.

A further example is that of sickle-cell anaemia. As you may recall, it is found mainly in people from Africa and their descendants. People who are homozygous for the sickle-cell gene have a defect in their haemoglobin: two of the protein chains that form part of the haemoglobin molecule have one amino acid replaced by another. When the haemoglobin is carrying oxygen it behaves normally, but after giving up oxygen to the tissues, the molecules may change shape so that they pack together to form conglomerations that distort the red blood cells. Instead of being disc-shaped, they become sickle-shaped (Figure 12.2).

People with sickle-cell anaemia are prone to both anaemia and sickling crises. Anaemia occurs because the red blood cells are destroyed prematurely by the body, so that there are fewer cells circulating in the blood than in a 'normal' person. They may also suffer from bouts of jaundice — the yellow colour of the skin in jaundice is due to large amounts of the breakdown products of haemoglobin circulating in the blood as a result of the fragile cells being rapidly destroyed. In the second major problem, sickling crises, large numbers of red blood cells change to the sickle shape. This phenomenon usually occurs when the oxygen level in the body is reduced, as may happen after strenuous exercise, during bouts of infection.

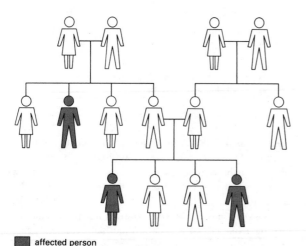

■ affected person

Figure 12.1 Pedigree of cystic fibrosis (an inherited disease caused by a recessive gene on an autosomal chromosome).

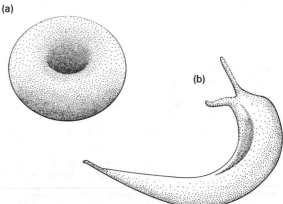

Figure 12.2 (a) A normal disc-shaped red blood cell and (b) one that has sickled as a result of containing sickle haemoglobin.

during pregnancy, or under stress. The sickled cells cannot flow normally through the blood vessels and often get stuck, causing obstructions. Not only may such crises be very painful, but the blockages can cause damage to various organs, particularly the eyes, kidneys and brain. The combination of anaemia and tissue damage may sometimes lead to death during childhood.

The gene for sickle-cell anaemia is not *completely* recessive. People who are heterozygous for the gene form both normal and abnormal haemoglobin. Though they show no signs of disease, their red blood cells will become sickle-shaped if oxygen levels in the blood are very low.

So far we have largely described gene defects that cause obvious diseases, regardless of the environment in which the person exists. Some genes, however, cause disease only under certain environmental conditions. An example is a disease known as phenylketonuria. People who are homozygous for the phenylketonuria gene have a defect in an enzyme in the liver. This enzyme usually catalyses the breakdown of phenylalanine, an amino acid contained in the diet, to another amino acid, tyrosine. The faulty enzyme allows large amounts of phenylalanine to accumulate in the body and cause damage to the brain, but damage occurs only if the diet contains phenylalanine. Therefore, babies in many countries, including the UK, are screened for the condition at the age of eight days, by examining their urine or blood for high levels of phenylalanine. If they have the disorder, they are put on a phenylalanine-free diet, thus preventing the brain damage and mental retardation that would otherwise occur. It is interesting that the defective gene for phenylketonuria is, like that for sickle haemoglobin, not completely recessive. People who have only one defective gene have slightly raised levels of phenylalanine in their blood, but not enough to cause any brain damage.

All the gene defects considered so far occur in autosomal chromosome pairs (that is, all pairs except the sex chromosomes). Disorders carried by recessive genes on the autosomes tend to appear 'out of the blue', since the parents are not aware that they carry the defective gene. If the defective gene is carried on the sex chromosomes, however, the parents may be aware of a family history of a particular disease. The X- and Y-chromosomes are of different lengths and carry genes for different proteins (though they do share some genes). If a recessive gene is present on one X-chromosome in a woman, then it will not code for its protein if there is a dominant gene on the other X-chromosome to suppress its action. In contrast, a recessive gene on the X-chromosome in a man *will* code for a protein, because there is no second X-chromosome. Similarly, a recessive gene on the Y-chromosome will also code for its protein. Therefore, some defects are specific to one or the other sex, but males are more likely to suffer from them than females.

Figure 12.3 shows the pedigree of some of the descendants of Queen Victoria, some of whom suffered from haemophilia. In this disorder, one of the enzymes involved in blood clotting is defective, so that spontaneous bleeds into the joints and muscles tend to occur.

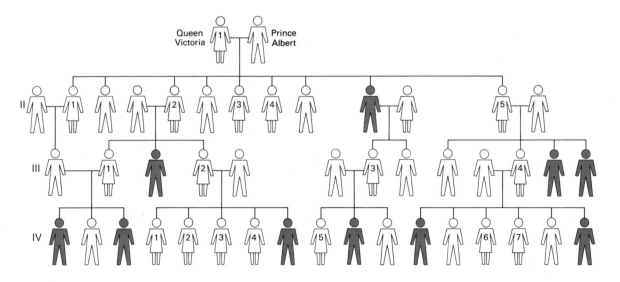

■ affected person

Figure 12.3 Pedigree of some of the descendants of Queen Victoria showing incidence of haemophilia, with her female descendants numbered in lines II to IV. Among the affected descendants were Leopold, Duke of Albany; Prince Maurice of Battenburg; Waldemar of Prussia; Tsarevitch Alexis of Russia; and Alfonso of Spain.

□ This disorder is caused by a sex-linked recessive gene* and occurs mostly in males. Study the pedigree and try to work out whether it is carried on the X- or the Y-chromosome?

■ The X-chromosome. You can tell this because it is inherited through the mother each time. Since women have two X-chromosomes, their sons (who are XY) must get their X-chromosome from their mother.

□ Which women in the pedigree *must* be carriers of the gene?

■ I 1; II 2 and 5; III 1, 2, 3 and 4. All these women had affected sons, although other women in the pedigree *may* also be carriers.

Interestingly, there was no history of haemophilia in the family before this. It would seem it must have arisen as a mutation when Queen Victoria was conceived.

Having looked at some of the diseases caused by single-gene defects, we now turn our attention to diseases that arise from disorders affecting a whole chromosome.

Chromosomal disorders

Chromosomal disorders are of two main types: abnormalities within chromosomes and abnormalities in the number of chromosomes. The former arise when breaks in chromosomes, which normally occur quite frequently, are incorrectly repaired. A broken piece may be lost, or may be re-attached to the 'wrong' chromosome, known as a *translocation*. These errors occur randomly and so cause a wide variety of disorders, each of which is rare. In this chapter we shall confine our discussion to the second type of chromosome disorder: abnormalities in the number of chromosomes.

□ How do you think this might occur? (Think back to what happens to the chromosomes during meiosis.)

■ During meiosis each cell divides, leaving only one chromosome of each of the twenty-three pairs in the daughter cells (ova or sperm). Sometimes an error occurs during this process and a pair of chromosomes fails to split.

This phenomenon, known as *non-disjunction*, results in one daughter cell having an extra chromosome and the other one lacking a chromosome. Down's syndrome, which used to be called mongolism, usually arises as a result of non-disjunction of chromosome number 21 during the formation of a sperm or ovum. Five per cent of cases, however, arise from translocation. Figure 12.4 shows the chromosomes of a person with Down's syndrome. It is not known why an extra chromosome-21 (known as trisomy-21) should cause the typical appearance of children with

* Some forms of haemophilia are not sex-linked: for example, Von Willibrand's disease.

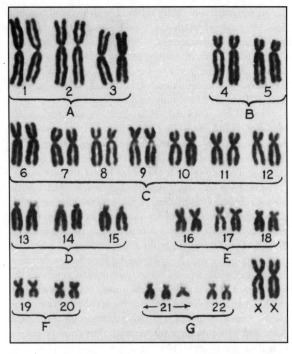

Figure 12.4 Chromosomes from a person with Down's syndrome — trisomy-21. Note that chromosomes can be seen only during cell division. Thus in the figure each chromosome appears as two strands joined together in preparation for the creation of two daughter cells.

Figure 12.5 The characteristic appearance of a Down's syndrome child.

Down's syndrome (figure 12.5), nor why people with Down's syndrome tend to be mentally retarded, below average height and sometimes have structural abnormalities of the heart.

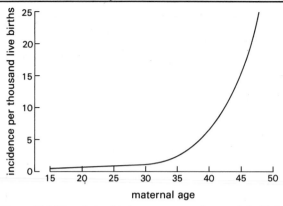

Figure 12.6 The relationship between maternal age and the birth of children with Down's syndrome.

The incidence of Down's syndrome in the UK is about 1 in 400 births. However, the incidence of cases resulting from non-disjunction varies with maternal age, being commonest, for unknown reasons, in older mothers (Figure 12.6). For this reason older mothers may be offered a screening technique (known as amniocentesis) during early pregnancy. A small amount of the amniotic fluid that surrounds the fetus is removed and stray fetal cells, present in all pregnancies, are examined for chromosomal abnormalities (including that of trisomy-21). If such abnormalities are found, the mother is offered the option of having the pregnancy terminated.

Abnormalities also occur in the number of sex chromosomes, caused by non-disjunction of the X- or Y-chromosome during cell division. Various abnormalities in the number of sex chromosomes have been found. Two examples are XXY (frequency 1 in 500 live births) and XO (frequency 1 or 2 in 5 000 live births). People with XXY chromosomes (two Xs and a Y) develop as men with testes and a penis. However, their testes are small, they are infertile, and their breasts develop. People lacking a second sex chromosome (XO) have female genitals. They tend to be short with developmental abnormalities. They also have abnormal ovaries that do not secrete the oestrogen necessary to develop secondary sexual characteristics.

☐ The chromosomal abnormalities described above include two of the more common abnormalities of chromosome number. Can you think why it is extremely rare to find babies with increased numbers of the other chromosomes?

■ Other abnormalities do occur, but they seem to cause such severe damage that the fetus is not viable and dies in early pregnancy, leading to spontaneous abortion. Examination of fetuses which have been spontaneously aborted has shown the presence of most of the theoretically possible trisomies, such as an extra chromosome 8, 13, 18 or 22.

Genetic susceptibility to disease

In Chapter 11 we introduced the idea of proximate and distal causes of disease. So far in this chapter we have discussed some examples of diseases in which the proximate cause is genetic, either a single-gene defect or a chromosomal abnormality. These form only a tiny proportion of the total number of diseases that occur. However, that is not to say that genetics does not play a part in the occurrence of all other diseases. Indeed, people's 'constitution' appears to be highly influential in determining whether or not they fall ill and with which disease. In other words, genetic make-up is a distal cause of many diseases.

This genetic influence is thought to result from several genes rather than a single one, such as in sickle haemoglobin. People vary, for instance, in such aspects as their immune responsiveness, their levels of hormone production, their metabolism of fats (such as cholesterol) and many other physiological processes. Given that people differ genetically in all these ways, particular environments produce different effects in different people. For instance, bacteria that cause tuberculosis may be acquired by several people, but only a few may fall ill with the disease. This would be partly a consequence of the latter group having a less efficient immune system than the others. Similarly, it is believed that a given diet may result in diabetes in some people, but not in others, as a consequence of variation in metabolism. While it would clearly be misleading to refer to influenza as a 'genetic' disease (in the same sense as we described sickle-cell disease), it is important to appreciate that genetic factors do influence its occurrence.

The next three chapters are devoted to diseases in which the proximate cause is quite clearly outside the body, in the environment. However, in Chapter 16, we shall return to this question of genetic factors in our discussion of diseases of unknown cause, diseases for which there are, as yet, no recognised proximate causes although several distal ones, including genetic make-up, are known.

Objectives for Chapter 12

When you have studied this chapter, you should be able to:

12.1 Give examples of the different ways in which abnormalities may arise in genes or whole chromosomes and know what factors determine whether they cause disease.

12.2 Explain how dominant-, recessive- and sex-linked-recessive gene defects are inherited.

Questions for Chapter 12

1 (*Objective 12.1*) How may an abnormality in just one protein manufactured by a cell cause disease, when it is only one among thousands of other proteins that are normal and functioning?

2 (*Objective 12.2*) Figure 12.7 shows a family tree where some members suffered from colour blindness. What can you say about the gene that causes this condition?

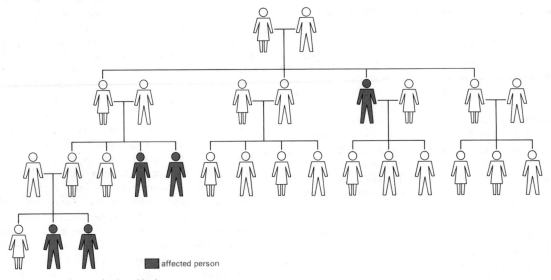

affected person

Figure 12.7 Pedigree of colour blindness.

13

Living with the physical world

This chapter includes material on nutritional diseases, building on what you have read in Chapter 12 of Book III, *The Health of Nations*, and in the article by Irvine Loudon, 'The history of pernicious anaemia from 1822 to the present day' in the Course Reader (Part 1, Section 1.6).

In this chapter we shall look at the interaction of humans with their physical environment and the ways in which diseases can result from environmental factors. In considering this subject, it is important to remember, as has been stressed before, the complex ways in which the human body interacts with the external environment while maintaining its own internal homeostasis. The body is not a solid lump of inert material — it continually takes in substances from outside and excretes substances it does not need. Inside the body, proteins are made constantly to replace worn out structures and enzymes. Enzymes catalyse and control complex step-by-step chain reactions that release or store energy or form chemicals needed by the body. Far from being static, the body can be imagined as a seething mass of chemical reactions maintained in a fine balance. It is on this complex system that outside factors operate and it is this system that they sometimes disturb.

There are two main categories of factors in the physical environment that affect health: animate and inanimate. In Chapter 14, the many and varied animate factors, such as viruses, bacteria, protozoa and helminths, will be discussed. Here we shall consider the inanimate ones. The inanimate physical environment, of course, varies between different places. The inhabitants of an Andean village in Peru experience quite different environmental conditions from those in an industrial town in the UK.

☐ In what ways do you think these two environments will differ in terms of such factors as climate, altitude, food, noise, chemicals and radiation?

■ *Climate* The Peruvian village, being near the equator, receives long hours of sunshine, though its altitude means it is cold at night. The British town has less sun and a wetter, temperate climate.

Altitude This will affect both the amount of oxygen available and the level of radiation received from the sun.

Captain James Lind, author of *A Treatise on the Scurvy*, 1753.

Table 13.1 Some vitamin-deficiency diseases

Vitamin	Source	Effect of deficiency
A (retinol)	dairy products, liver	night blindness, thickened skin, corneal damage
B1 (thiamine)	whole cereal grains, pulses	beri-beri
nicotinic acid (niacin)	meal, fish, yeast extract, milk	pellagra
B12 (cobalamin)	liver (if undercooked), other animal products	disease of spinal cord, anaemia
C (ascorbic acid)	fresh fruit and vegetables	scurvy
D (cholecalciferol)	dairy products, fish liver oils, fortified margarine, manufactured in skin in presence of ultraviolet light	rickets, osteomalacia
folic acid	yeast extracts, green vegetables, liver	anaemia

Food Both the amount and type of food consumed will differ.

Noise The Andean village will be much quieter, since there are few, if any, cars and powered machinery.

Chemicals People in the Andean village may be exposed to naturally occurring harmful chemicals in plants and animals, whereas in the UK the source of chemicals is more likely to be polluted air, occupational environments and intensively farmed foodstuffs.

Radiation There will be variations in the natural background levels of radiation from the rocks on which the towns are built. In the UK, there are additional sources of radiation (such as medical uses).

It would appear, therefore, that some of the international variation in the spectrum of disease has its origin in differences in physical environments. We are going to concentrate on six main factors: food, physical trauma, extremes of temperature, chemicals, hypersensitivity (allergy), and radiation (including noise). The importance and influence of both water and oxygen on human health have already been discussed in Chapter 8.

Food

You have already learnt a fair amount about people's need for an adequate diet and the effects of deficiencies of protein, carbohydrate, vitamins and trace metals in the diet.* We shall consider only some specific deficiency diseases and obesity in this chapter.

* This is dealt with in Chapter 12 of *The Health of Nations* (U205, Book III).

□ Do you remember what vitamins are?

■ The term vitamin is used to cover various chemical compounds that are needed in small amounts by the body, usually to make essential enzymes or co-enzymes, substances which activate enzymes. The vitamins are called A, B, C, etc. because these were the names they were given as they were discovered during the early years of this century, before their chemical structures were known. The so-called B-vitamin turned out to comprise a large number of related substances with quite separate biochemical effects, now designated vitamin B1, B2, etc.

Table 13.1 shows the sources of some vitamins whose lack is known to cause deficiency diseases in humans. We shall start by describing two of the main diseases due to vitamin deficiencies: rickets and keratomalacia.

Vitamin D is important in the deposition of calcium in bones. In adults, a lack of vitamin D leads to thin bones which fracture easily (osteomalacia). In growing children, deformities of the limbs may occur as a result of bowing of the softened bones, a condition known as rickets (Figure 13.1). In the UK, the main dietary sources of vitamin D are margarine (to which vitamin D is added during manufacture), fish, eggs and to a lesser extent, dairy products. However, it is also formed in the skin under the influence of ultraviolet light from the sun. In dark-skinned people ultraviolet light does not penetrate the skin so easily. This means that in countries without much sun, dark-skinned people are at risk of developing vitamin D deficiency if their dietary intake is inadequate.

Vitamin A deficiency is a major cause of blindness in

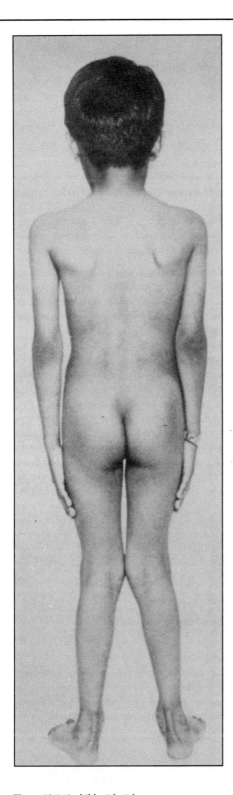

Figure 13.1 A child with rickets.

children in countries where nutrition is poor. The main abnormality occurs in the horny layer of the skin, which fails to be shed from the skin surface in the usual way. As a result, the skin becomes hard and thickened, a condition known as keratomalacia. When this occurs on the transparent covering of the eye (the cornea), it becomes thickened, hazy and dry. This dryness makes it susceptible to infections which cause further damage. The cornea may ulcerate and scar or it may even split completely. Both these changes lead to blindness.

Apart from osteomalacia in elderly women, thought to result from an inadequate dietary intake and insufficient exposure to sunlight, these vitamin deficiencies are rarely seen in the UK. One that *does* occur relatively frequently is B12 deficiency. It usually occurs in people who have defective *absorption* of B12 rather than an inadequate diet.

☐   Do you know the two damaging effects of vitamin B12 deficiency?
■   A form of anaemia, known as pernicious anaemia, and damage to the nerves in the spinal cord.

Anaemia can also result from a deficient intake of folic acid, a chemical that is usually classified with vitamins. The commonest cause of anaemia in the UK, however, is not the lack of a vitamin, but the lack of iron. Metals are rarely lacking in the diet, most of them being required in only minute amounts, iron and iodine being the only two that are commonly in short supply.

Iron is needed in the synthesis of haemoglobin and its deficiency means that fewer red blood cells are made. If iron is lacking, the red blood cells that are made are small and contain a low concentration of haemoglobin. With a fall in the level of haemoglobin the oxygen-carrying capacity of the blood is reduced and may be inadequate for any exertion. Activities such as climbing the stairs may lead to breathlessness when the haemoglobin level is about half the normal level.

Men and non-menstruating women need about 1 milligram of iron a day to replace the normal losses from the body (in dead skin cells, for instance). Menstruating women need twice as much as this because they lose iron in menstrual blood. Most 'good' diets provide 12–15 milligrams of iron a day, but only about 2 mg of this is absorbable. It is common for women with very heavy periods to have haemoglobin levels in the blood which are only 80 per cent of the average level. However, they do not usually feel ill because of this, and there is some controversy as to whether such levels should be considered as 'abnormal' and needing treatment.

Iodine deficiency is still common in many countries and used to be common in upland areas of the UK, for example Derbyshire, where the iodine content of the soil is low. Iodine is used by the thyroid gland in making thyroxine, the hormone that regulates metabolism.

☐ As you may recall, the thyroid gland is controlled by the pituitary gland. What do you think the effect of insufficient iodine in the diet will be?

■ Insufficient iodine results in the thyroid manufacturing inadequate amounts of thyroxine. Low blood levels of thyroxine will result in the pituitary secreting more thyroid-stimulating hormone. This has the effect of making the thyroid enlarge in an attempt to increase thyroxine production.

The resulting swelling is known as a goitre (Figure 13.2). The swelling does not usually cause any harm, but the low levels of thyroxine may do so, particularly in pregnant women and children. Severe iodine deficiency during fetal life and childhood impairs growth and retards the development of the nervous system (in particular, causing deafness). This condition is known as cretinism, and is still common in mountainous areas of South America, Asia and Australasia, where iodine deficiency is prevalent as a result of low concentrations in rainfall.

In industrialised societies, ill-health due to dietary deficiencies is overshadowed by the ill-health associated with over-consumption. The consumption of more food than is necessary may result in obesity. The rate at which people use energy varies — some people can eat more than others and not get fat. The reasons for these differences are only just beginning to be understood. Some of them result from slight variations in people's metabolic pathways. Several studies have shown that fat people tend to take much less exercise than thin people, even when

Figure 13.2 Goitre — an enlarged thyroid gland due to a lack of iodine in the diet.

engaged in the same activity. In one study in the USA, fat people and thin people in similar occupations wore pedometers (instruments for measuring the distance walked) for one week. The thin people walked on average two-and-a-half times as far in a week as the fat ones. Whether the fat people were overweight *because* they took less exercise, or whether they walked less because they were fat, remains uncertain.

☐ Being obese is not good for health. From your general knowledge and reading so far, do you know which diseases account for the higher mortality in obese people?

■ The principal ones are diabetes, coronary heart disease, hypertension (raised blood pressure), strokes, and diseases of the gall-bladder and large bowel (colon).

There is considerable additional morbidity associated with obesity. This includes joint problems, such as osteoarthrosis of the hips and knees, due to the increased load the joints have to bear; breathlessness and respiratory infections; varicose veins; psychological problems.

The possible harmful effects of the type of diet consumed in industrialised societies is not confined to the amount of food; the type of food is also important. In recent years most attention has focused on the amount of fibre in the diet and the amount and type of fat consumed.

The benefits of a diet with a *high* fibre content have been deduced from comparisons of East African rural populations on a high-fibre diet with those in cities on a low-fibre diet. In people eating a lot of cereal fibre, food and faeces travel rapidly through the intestine, and the faeces are bulky. A low-fibre diet has been said to be responsible for inflammation of the appendix (appendicitis), inflammation of the large intestine (diverticulitis) and cancer of the large intestine, as well as obesity and diabetes. Although it is true that people on high-fibre diets rarely suffer from these diseases, you will realise by now that in comparing different populations there are many variables other than food which must be considered. Nevertheless, there is general agreement that a moderate intake of fibre lessens the risk of developing some of these diseases.

Physical trauma

The human body can withstand only a certain amount of physical shock (trauma). In the very short-term, survival after an accident depends on two factors — the ability to breathe and the integrity of the circulatory system.

☐ Why do you think these two factors should be so important?

■ Cells need a constant supply of oxygen. Breathing ensures a supply of oxygen, which the circulatory system carries to the cells.

The commonest reasons for dying as a result of trauma are, therefore, loss of blood, damage to the airways, or brain damage which interferes with the control of breathing. If the immediate damage is survived, then the body begins to repair itself in two main ways: by *cell division*, and by the formation of *scar tissue*, made up from collagen fibres. The relative contributions of these two processes depends on the site of the damage and the conditions around the damaged area. Not all cells retain the capacity to divide that they had during fetal life — muscle and nerve cells, for instance. Repair of these tissues is therefore entirely by the deposition of collagen. In contrast, bone heals almost entirely by the formation of new cells. When a bone is broken, the space between the two broken ends becomes filled with a blood clot. Then cells from the surface layer of the bone creep into the space, where they settle and secrete the connective tissue basis of bone. Finally, calcium salts are deposited to form new bone. The final result is indistinguishable from the old bone (if there is no deformity).

Skin comes somewhere between these extremes. Damage to the epidermis (the outer layer) is repaired perfectly by new cells growing up from the deeper cells (which divide continuously throughout life). If the damage penetrates right through to the dermis then, although cells migrate in from the edges, collagen is laid down and, as a consequence, the repair leaves a scar.

The process of cellular repair in any tissue is dependent on a good blood supply. Thus, old people with a poor circulation in the skin on the legs frequently have a diminished capacity to heal. A small cut or graze on the shin or ankle may, instead of healing, develop into an ulcer, an area where the surface covering is missing and the tissue beneath is exposed. Figure 13.3 shows a severe example of such a non-healing ulcer.

When healing is achieved by scar tissue, the resulting repair is strong. Scar tissue, however, does have some disadvantages compared with normal tissue. In particular,

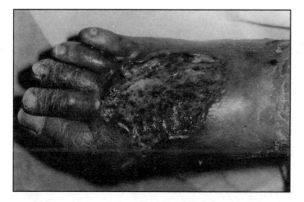

Figure 13.3 Skin ulcers on the leg caused by poor circulation.

it has the disadvantage that it shrinks with time and this may cause a local deformity: for instance, scarring in the oesophagus (the gullet) may cause it to narrow, so that swallowing food is difficult. Similarly, scar tissue in the brain may contract and pull on adjacent neurons. This 'pulling' may make the cells more likely to initiate nerve impulses. Such abnormal excitability in a group of neurons is one cause of epileptic fits, which occur when the synchronous 'firing' of a small group of nerve cells spreads to other nerve cells throughout the brain.

Extremes of temperature

You saw in Chapter 8 how the body controls its own temperature at a constant level (about 37 °C), despite changes in the temperature of its surroundings. At extreme temperatures, both very cold and very hot, the body may be unable to maintain control and damage may occur.

If the cells of the skin are subjected to temperatures above 45 °C for more than a minute the nuclei within the cells begin to swell. Above 52 °C the cells are irreversibly damaged: they swell, their nuclei become misshapen, and they may become separated from the tissue beneath them, as happens when a blister forms. Above 60 °C the proteins within the cells start to coagulate: that is, they change in the same way as the white of an egg does when it is boiled. This means that none of the enzymes in the cell can function any more — the cell is well and truly dead by this time. The release of cell contents stimulates an acute inflammatory response in the skin and it is this that causes the redness and pain associated with burns. The way in which a burn heals depends on the extent of the damage. If only the epidermal cells of the skin are damaged, the skin can heal from its deeper cells, in a similar way to superficial cuts. If the damage includes the dermis, then the skin has to grow in from the undamaged edges of the burn, a process which may take months if a large area of skin is involved. Collagen will also be laid down in the healing process. The shrinkage of collagen following the healing of large burns may cause considerable disability, because the skin becomes shrunken and rigid and may impair the movement of joints.

Frostbite occurs when a part of the body becomes frozen, as can occur if the surrounding temperature is less than 0 °C. The commonest sites affected are the fingers, toes, nose and ears, all of which have a large surface area through which to lose heat. The first sign of frostbite is numbness and extreme whiteness of the extremity, though later the skin may appear bluish. The skin may be hard and brittle because the tissues beneath it are frozen solid. The tissues are damaged by ice crystals, and by the effects of being without oxygen because the blood flow is stopped. After the tissue is warmed up again, some of it may be so damaged that it fails to recover. Healing will once again be by cell division and scar tissue.

Chemicals

It is known that some chemicals cause disease.

☐ From your reading about the biochemistry of the body, can you suggest why this might be so?

■ The body is maintained by complex pathways of chemical reactions which culminate either in the liberation of energy or in the production of a chemical that is needed for growth or some other function. These metabolic pathways will function normally so long as the internal environment of the body remains stable. New or uncommon chemical substances may disrupt these pathways, by reacting with an enzyme or an intermediate product and preventing the next stage in the sequence.

This disruption of a chemical pathway prevents the formation of its end product, and may cause the build up of an intermediate product which then interferes with other reactions. You have already seen in Chapter 12 how the absence of a single enzyme may cause widespread disruptive effects in the body. Chemicals which seriously disrupt normal working are known as poisons. Some of these occur naturally and have been well-known for centuries. The Israelites in the wilderness ate quails and were immediately smitten by a 'plague' (Numbers 11:31–33) — which is strange, since quail are usually good to eat. However, when quail migrate from Africa to Europe they feed on hemlock seed, fine for quail, but poisonous to humans, who suffer nausea, vomiting and spreading paralysis after their meal.

☐ What poisons occur naturally in the UK?

■ The more common ones are listed in Table 13.2.

The range of naturally occurring poisons pales into insignificance beside the products of industry. Figure 13.4 shows some of the range of chemicals that are manufactured, the vast majority of which can cause harm to humans

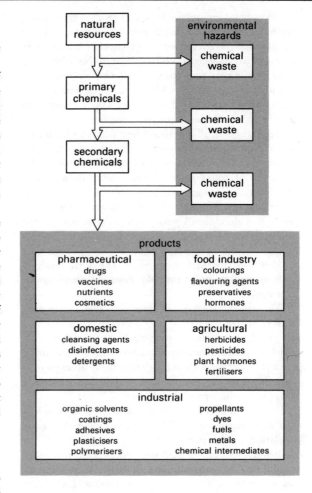

Figure 13.4 The production and use of chemicals.

Table 13.2 Some of the poisonous plants and animals found in the UK.

Fungi:	
Fly-agaric	rarely causes death.
Death-cap	gives symptoms 6–15 hours after ingestion. Death occurs in 90% of cases.
Green plants:	
Yew	one of the most poisonous plants in the UK, though cases of poisoning are rare, perhaps because its toxicity is well-known. Yew wood was used to make long-bows and yews were often grown in church yards, which were enclosed, so that cattle could not feed on the leaves and berries.
Laburnum	grown in paths and gardens. Laburnum seeds are one of the commonest causes of serious poisoning in the UK, but fatalities are rare.
Fox-glove	contains digitalis, which is used in the treatment of heart disease, but causes disturbances in the heart in overdose.
Deadly-nightshade	glossy black berries the size of cherries which are very attractive to children. 20–30 berries may cause death.
Animals:	
Honey-bee and wasp	both stings cause local pain and inflammation.
European viper (the adder)	venom usually causes localised swelling and pain, but death rarely occurs.

if not used as intended. In addition, during each stage of their manufacture, waste chemicals are created and released into the air, rivers and seas. This is not the end of the story: the waste chemicals may be taken up by plants and animals, and may finally be eaten by humans. For example, mercury-containing waste has been frequently dumped into the sea. There, microorganisms ingest it and convert it to compounds that are even more toxic than the original ones. The microorganisms are eaten by small fish, which in turn are food for swordfish, tuna and cod. If eaten in large quantities, these compounds cause damage to the nervous system in humans. The problem is particularly acute in Japan, where fish is eaten in large quantities.

Another chemical that can enter the food chain is lead, which is given out in exhaust fumes of cars burning lead-containing petrol and deposited on the soil near roads. Vegetables grown on such soil contain high concentrations of lead. The body is not able to excrete the lead ingested through eating contaminated food and as a result the metal accumulates in the body. High levels of lead cause widespread damage by disrupting protein molecules involved both in cell structures and as enzymes. There is considerable evidence that a high level of lead in children retards their mental development.

The use of insecticides and herbicides has also created environmental health hazards. One of the most notorious of these is a herbicide, code-named 2,4,5-T, that was produced in bulk in the 1960s and early 1970s for use in agricultural and forestry work. It was also used by the US army in Indochina, where it was sprayed from planes on to crops, to drive peasants off the land into the villages, and on forests, to destroy guerilla cover. 2,4,5-T was supposed to be poisonous only to plants, but the Vietnamese began to report symptoms in animals and humans. Fish and farm animals died and people who had been in the path of the sprays, or had eaten food which had been sprayed, complained of nausea, lassitude and persistent skin rashes of a type known as chloracne. By 1970 there were reports that women who had been in the sprayed area were giving birth to malformed babies. These reports were originally discounted in the West, but it then became apparent that the 2,4,5-T was contaminated with a by-product generated during its factory synthesis, the substance dioxin. Dioxin is extremely toxic — a few grams would be sufficient to wipe out the population of London. In sub-lethal doses it has the effects reported by the Vietnamese and is also now known to be carcinogenic (cause cancer). American veterans of the Vietnam war were awarded $180 million in damages in 1984 against the manufacturers of 2,4,5-T as compensation for the long-term effects of the chemical.

There are three main ways in which chemicals cause damage to cells: by damaging cell structure; by interfering with chemical reactions; and by blocking nerve transmission. Cell structure may be damaged in a number of ways. For instance, strong acids on the skin change the structure of protein molecules, thereby causing cell death; solvents and detergents dissolve the fats that are important in maintaining the softness and suppleness of skin, thus causing dryness and cracking of the skin. Carbon monoxide, a chemical produced by the partial combustion of petrol or domestic gas, combines with haemoglobin in red blood cells so that the haemoglobin molecule can no longer carry oxygen around the body. Death results if the dose of carbon monoxide is sufficient to combine with a large proportion of the haemoglobin in the blood. Many of the military nerve gases work by blocking the receptor sites on cells that respond to neurotransmitters (chemicals released at nerve endings). Similarly, tubocurare, a paralysing poison used by South American Indians in hunting, blocks receptors on muscle cells. The animal is unable to escape because its muscles will not contract. It is safe for the Indians to eat an animal that has been poisoned in this way because tubocurare is not absorbed through the gut — to have effect, it must be injected into the blood.

Some chemicals cause harm only if taken in high doses. This is true of many synthetic drugs, such as aspirin, that are marketed for their beneficial actions on the body. Paracetamol can cause liver damage and may lead to jaundice and death a few days after an overdose.

☐ Why do you think that many drugs and poisons have their most damaging effect on the liver, rather than any other organ?
■ The liver is responsible for rendering many toxic molecules harmless. Therefore high concentrations of the substances may build up in that organ.

Chemical damage need not be acute and sudden. Continual small doses of a poison may cause gradual cell damage and scarring. Alcohol is mildly poisonous to liver cells. If excessive amounts of alcohol are taken over a long time, the cells die and become replaced by scar tissue. This is known as cirrhosis of the liver and is also seen in other kinds of chronic poisoning, for instance, with carbon tetrachloride (dry-cleaning fluid) and methylated spirit.

Various occupations expose workers to particular chemical 'poisons'. Silica particles, which cause chronic lung disease, are inhaled by sand-blasters and stone masons. The silica particles are ingested by macrophages, but the particles have a damaging effect on the lysosome membranes inside these cells: the lysosomes release the enzymes that they contain, thus killing the macrophage. The lysosomal enzymes are finally released into the lung tissue. The resulting damage heals with scarring, which may eventually become extensive enough to impair lung function — a disease known as silicosis.

Silicosis is an uncommon disease in the general population compared with chronic bronchitis which affects between 8 and 17 per cent of men in the UK. In this condition there is excessive mucus production in the lungs in response to inhaled irritants. This excess mucus tends to block the smaller bronchi which predisposes to infections, which further damage the lungs by narrowing the smallest bronchi. Sometimes, in addition, the walls of the alveoli also break down, a complication which is known as emphysema. In its early stages chronic bronchitis causes a cough in the morning and the need to bring up mucus. Once the architecture of the lung begins to change, breathing may become difficult until finally insufficient oxygen can be taken in to maintain life. The prevalence of chronic bronchitis is directly related to the levels of atmospheric pollution and is aggravated by smoking. It used to be called 'the English disease' because it was almost entirely associated with industrialisation. It is now strongly associated with smoking.

Emphysema can occur in young adults without its usual association with chronic bronchitis. Again, it occurs only in areas with atmospheric pollution or in smokers, but is confined to people who have a genetic predisposition for the disease. Potential sufferers have a defect in a chemical that inactivates an enzyme released by white blood cells in response to inhaled bacterial or chemical irritants. Normally the enzyme is inactivated in a short time, otherwise it catalyses the breakdown of protein in the lungs, particularly of elastin, the protein which gives elasticity to structures in the lung. The alveoli in the lung become more rigid and breathing becomes extremely difficult. This type of emphysema provides a good example of the interaction of genetic with environmental factors in the cause of disease: the environmental irritant means that the destructive enzyme is released and the genetic lack of inactivator means that the destructive enzyme has widespread effects.

Some chemicals cause damage to DNA molecules within the cell and this can lead to the growth of cancers. In 1775 Percival Potts was the first person to describe a chemically induced cancer when he noted that chimney sweeps had a high incidence of cancer of the scrotum and suggested that this was due to the action of soot. Since then, several chemicals have been implicated in the development of cancers. Oils and tars may cause skin cancer in people whose skin is regularly exposed to these chemicals and it is thought to be the oils and tars in cigarette smoke that lead to the development of lung cancer. The association between cigarette smoking and lung cancer is now well established. It is estimated that the incidence of lung cancer would drop by 90 per cent if no one smoked.

Cancer of the lung also occurs in people who have been exposed to asbestos. More rarely, they develop cancer of the epithelium that covers the surface of the lung. This type of cancer, known as mesothelioma, is practically never found in the absence of asbestos fibres in the lung. Inhaled asbestos does not *always* cause cancer, but it may produce changes similar to those found in silicosis and chronic bronchitis.

Certain dyes may cause cancer. Once taken into the body, these dyes are excreted in the urine. The incidence of cancer of the bladder in people who work in the manufacture of aromatic dyes is twenty-five times that in the rest of the population.

Chemicals ingested or absorbed by a pregnant woman may cause abnormalities of the fetus even though they do not harm the mother. They are known as *teratogens* (literally, monster forming).

☐ At what stages of development might chemicals damage a fetus?
■ They might affect the DNA at an early stage, causing widespread genetic effects, or they might interfere with the biochemical pathways involved in cell division as the fetus develops.

Drugs are the major cause of such disturbance. The most notorious case of drug-induced deformity is thalidomide, marketed as a remedy for morning sickness in early pregnancy, but later found to cause abnormalities in the growth of limbs. Many other drugs are known to cause fetal abnormalities and to affect the overall growth of the fetus.

Alcohol also crosses the placental barrier and is now believed to affect the fetus, even in small amounts. The most extreme effects are observed in the offspring of mothers who are alcoholics. These effects include retardation in growth and intelligence, and deformities of the head, face and joints, and have been categorised together as fetal alcohol syndrome.

The harmful effects discussed so far can occur in *everyone* who absorbs, inhales or comes into contact with these chemicals. There is, however, a second category of chemicals that result in harm only in *certain* people, who are then said to be hypersensitive, or allergic, to this or that chemical. Hypersensitivity is said to exist when the immune response to a non-self substance produces harmful effects, rather than protecting against the antigen. Adverse immune reactions are of several types. We shall look at only two: one mediated by circulating antibodies and one by T lymphocytes.

Circulating antibodies are implicated in hay fever, some types of asthma and severe reactions to wasp and bee stings. When the non-self material (known in this case as an allergen) enters the body, it binds to antibodies that are attached to the membrane of mast cells and causes those cells to release an excessive amount of histamine and other chemicals. It is these chemicals that produce the ·allergic reaction. This type of reaction particularly affects mucous membranes (for example, in the nasal passages and inside the eyelids) and skin, where mast cells abound. People with hay fever are usually allergic to pollen, which combines with antibodies attached to mast cells at the back of the nose. Asthma may be caused by allergens such as the mites that live in house dust. Swelling of the mucous membranes and spasm of the muscles in the walls causes narrowing of the bronchi and difficulty in breathing. The type of allergy found in asthma and hay fever tends to run in families and may, therefore, have some genetic basis.

The second type of hypersensitivity is mediated by T lymphocytes. This type of reaction can occur after skin contact with a chemical that combines with proteins in the skin to form an antigen. The resulting lymphocyte proliferation leads to skin redness, blistering and weeping, known as contact dermatitis. Initially, the reaction is localised in the region of the skin that has been in contact with the chemical, but later it may spread to other areas. The initial distribution may give a clue to the causative agent, as can be seen in the examples in Figure 13.5.

We should not leave this section on hypersensitivity without mentioning idiosyncratic reactions to drugs. All drugs cause adverse reactions in at least some people. The mechanism for these reactions is not always understood, but it is generally assumed to be an inappropriate immune response. One of the commonest types of reaction are skin rashes. Figure 13.6 shows an example. Although all drugs can cause such effects, some drugs do so more than others. Once someone has developed antibodies or lymphocytes sensitised to a particular drug, they will always react to it, and the effects usually worsen with each exposure. For example, individuals with an allergy to the antibiotic penicillin may react so severely to the drug after several exposures that it can prove fatal.

Radiation

Radiation and its effect on people's health is very much in the news at present. Questions ranging from the harmful effects of sunbathing and microwave ovens to the safety of nuclear power are prominent in the media. Both natural and artificial radiation' are part of the environment, but what is radiation exactly?

In Chapter 2 we explained that all chemical elements were made up of atoms, but we did not discuss the internal structure of the atoms. All atoms can be thought of as consisting of a central nucleus made up of subatomic particles called protons and neutrons.* In orbit around this

* The one naturally occurring exception is the hydrogen atom, the nucleus of which has just one proton and *no* neutron.

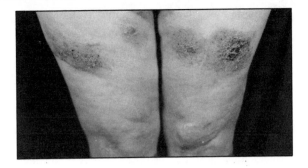

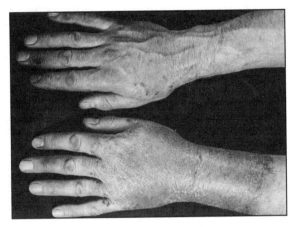

Figure 13.5 Contact dermatitis caused by (a) a nickel-containing suspender belt and (b) the use of a protective glove.

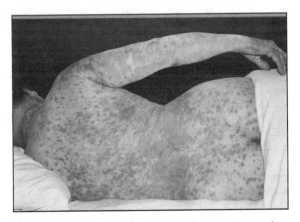

Figure 13.6 Skin reaction in a person hypersensitive to barbiturates.

central nucleus are a number of electrons. It is the electrons that are gained or lost when atoms become ions: for example, when sodium (Na) becomes the sodium ion (Na^+), it temporarily *loses* a negatively charged electron from orbit around its nucleus.

Atoms can give out energy, known as radiation, in two forms. The first, electromagnetic radiation, can be thought to consist either of weightless particles called photons, or of a series of electromagnetic waves that travel at a speed of 300 million metres per second. Electromagnetic radiation includes visible light, ultra-violet and infra-red radiation, radio waves and X-rays, as well as some of the radiation from radioactive materials. The second kind of radiation is in the form of heavier particles that are ejected from the nucleus, the most common being beta-particles (which are essentially the same as electrons), alpha-particles (which are made up of two protons and two neutrons) and neutrons.

◻ To what sources of radiation are humans exposed?
■ You might have thought of the Sun, other stars, radioactive elements in the Earth and the atmosphere, and artificial sources, such as X-rays and nuclear power installations. In fact, the air humans breathe, the food they eat and the tissues that compose their bodies are all slightly radioactive, that is, they all contain radioactive elements (elements whose nuclei spontaneously give out energy as particles).

Radiation is important because the energy it contains can be transferred to any object it meets. This energy transfer may be either useful or harmful.

◻ Do you know any examples of electromagnetic radiation that is useful to living organisms?
■ (i) Energy from visible light is absorbed by pigments in plant cells during photosynthesis and is used in the production of carbohydrates from carbon dioxide and water. (ii) Light energy is absorbed by cells at the back of the human eye — this forms the basis of the ability to see. (iii) Ultraviolet light is absorbed by human skin and is used there in the manufacture of vitamin D.

The harmful effects of radiation on biological tissues are not completely understood. Different radiations have different energies and therefore produce different effects. It is known that these effects depend on complicated processes of energy transfer to the biological molecules. This energy can disrupt the bonds between atoms in certain molecules: for instance, it can cause breaks in DNA and proteins. It can sometimes give *so much* energy to molecules that electrons leave the molecules, which are then said to be ionised. In this state the molecules are highly reactive and will combine with other chemicals in the vicinity, thus disrupting the normal chemical pathways in

the cell. This type of radiation is called ionising radiation and will be discussed more fully shortly. First, we shall consider the effects on health of the different types of non-ionising radiation.

Non-ionising radiation
This type of radiation includes microwaves, ultraviolet radiation and lasers. We shall consider each of these in turn.

Microwaves pass through the skin and are absorbed by water molecules in deep tissues, where the local heating causes burns. Since not much damage is done to the skin, which is the site of temperature receptors, there may be no warning sensation of burning and the damage caused may be extensive. The effect of low doses of microwaves on humans is not certain and safety regulations about the maximum permitted dose at work vary in different countries. It is known from experimental work on rabbits that microwaves can cause damage to the lens in the eye — the lens becomes cloudy and will no longer transmit light. The maximum permitted level for workers with microwaves in Britain is set at one tenth of the level known to cause this sort of damage to the eye.

Ultraviolet radiation is that part of the Sun's emission that causes sunburn and tanning. Ultraviolet light is particularly strongly absorbed by protein molecules and DNA in the cell, and, in humans, little or none passes through the skin. When exposure is excessive, the cells release substances that cause dilation of the blood vessels, with consequent reddening and warmth in the skin. Exposure to ultraviolet light stimulates the production of melanin, a brown pigment, in the skin. This is a useful safety phenomenon since the extra melanin tends to absorb the ultraviolet light and protect the other proteins in the cell.

The effect of ultraviolet light on DNA molecules can have long-term consequences. Breaks in DNA are usually repaired by a particular enzyme, but this mechanism is not always efficient and if mistakes occur in the repair, skin cancers may occur. The incidence of one type of skin cancer, known as melanoma, varies in people living at different latitudes.

◻ Why should the incidence of melanoma vary with latitude?
■ Exposure to ultraviolet light is greater nearer the equator. The resulting damage to DNA molecules may lead to cancer.

Ultraviolet light also affects the eye, causing inflammation of its surface. Such redness and pain occurs in snow blindness, from ultraviolet lamps, and in arc welders who do not wear adequate eye protection.

Lasers are devices which produce light consisting of only one wavelength, in a thin parallel beam of high intensity. Lasers can damage the eye if a person looks

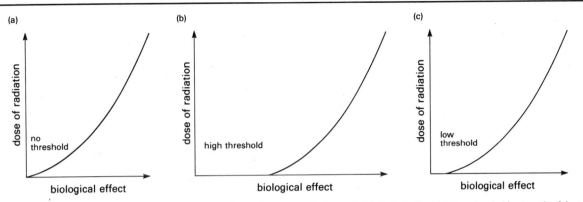

Figure 13.7 The possible relationships between low doses of ionising radiation and a biological effect: (a) if no threshold exists; (b) if there is a high threshold; and (c) if there is a low threshold.

directly at a laser beam as the light is focused by the lens of the eye on to a single spot on the retina. This concentration of light may burn the retina causing a permanent blind spot in a person's field of vision. Although such uncontrolled exposure may be harmful, lasers are used by doctors to treat conditions in which damage to the retina has occurred: for instance, to stop bleeding from retinal blood vessels in diabetes.

Ionising radiation

Ionising radiation, the kind that causes molecules to lose electrons, arouses the greatest public concern. Knowledge of its effects is not complete. Evidence for its harmful effects comes from four main sources: (i) studies of workers, who were exposed to radiation in industry before the dangers were known; (ii) animal experiments; (iii) studies of the effects of the atomic bomb explosions at Hiroshima and Nagasaki; and (iv) studies of people exposed to ionising radiation as part of their medical treatment (it used to be used in the treatment of a wide range of disorders but now it is used only in the treatment of cancers). The immediate effect of large amounts of ionising radiation is cell death. Different tissues show differing susceptibilities, the bone marrow being one of the most susceptible. In high whole-body radiation, the cells of the brain swell and the blood vessels to the brain become inflamed. If a person survives this stage, damage to the lining of the gut becomes evident with the onset of diarrhoea and vomiting — so-called radiation sickness. This may occur within hours or days depending on the severity of the exposure. Later, the effects on the bone marrow begin to show, as circulating cells in the blood are not replaced when they die. The loss of white cells leads to a high susceptibility to infections. Loss of platelets causes clotting disorders with spontaneous bleeds into the skin and internal organs.

Delayed effects have been shown by survivors of nuclear bomb explosions, who show an increase in the incidence of leukaemia (a cancer of white blood cells). This incidence is proportional to the amount of radiation received. Other cancers known to occur more often after radiation are those of the skin, bone, larynx, pharynx and lung. Leukaemia has also been seen in children who were irradiated while still in the uterus. Besides this tendency to induce cancers, radiation may cause cataracts in the eye, sterility, and chronic inflammation in the skin and blood vessels.

There may also be genetic effects. Damage to DNA can occur in sperm and ova, producing gene defects. This has been shown clearly by using high levels of exposure in experimental animals. An increase of gene defects in the children of atomic bomb survivors has not been demonstrated, but this is not really surprising — the majority of mutations that are passed on to children will be in recessive genes and may have no obvious effect until two people both carrying the same recessive gene have a child. The incidence of spontaneous abortion was high in survivors of Hiroshima and Nagasaki, which does suggest an increase in the number of lethal gene defects.

The dangers of ionising radiation that cause most concern, when considering the exposure of an average member of the population, are the risk of cancers and genetic defects. The effects of high doses are fairly clear, but what happens at low doses of radiation is still not certain. Figure 13.7 illustrates the problem.

If radiation levels have an adverse effect right down to zero levels (Figure 13.7a), then even very low levels of radiation will cause some harm. Alternatively, there may be a threshold level at which radiation starts to have effects. This threshold would indicate what was the level of safe doses of radiation (Figure 13.7 b and c). It is not known which of the three possibilities shown in Figure 13.7 is correct. This is one reason why defining 'safe' levels of radiation is difficult. If there is no threshold, then any level of radiation will cause an increase in the incidence of cancers. If there is a high threshold, then artificial radiation is probably safe as long as it is less than, or at the same sort of level, as natural radiation.

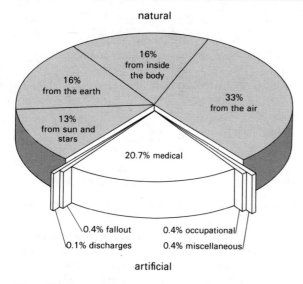

natural

16%
from inside
the body

16%
from the earth

33%
from the air

13%
from sun and
stars

20.7% medical

0.4% fallout   0.4% occupational
0.1% discharges   0.4% miscellaneous

artificial

Figure 13.8 Sources of ionising radiation in the UK based on the individual average annual dose.

Natural and artificial radiation levels

There is some agreement on the relative contributions of natural and artificial radiation (Figure 13.8).

☐ What proportion of radiation comes from artificial sources?

■ 22 per cent, most of which is from medical sources.

☐ Can you think of problems with such *average* figures?

■ Radiation, both natural and artificial, may not be distributed evenly.

For instance, some rocks contain more radioactive elements than others. In coastal areas of Kerala in Southern India and Guarapari in Brazil there are patches of sand that contain a high concentration of radioactive elements. This may increase the background levels of radiation to twenty times the average. In Britain igneous rocks and those with a high silica content have the highest radioactivity. If these rocks are used to build houses, the background levels of radiation within the houses may be up to ten times higher than average.

Artificial radiation is even less uniformly distributed. About one-third of the population have some sort of X-ray each year. The distribution of radioactivity from nuclear bomb testing and from discharges from nuclear power stations is not so well documented. Nuclear waste of low activity is discharged into the sea. Its dispersion depends on the tides and weather, and it may enter the food chains and become concentrated in fish. Even when this is taken into account, however, the estimates of doses to high-risk individuals, such as workers in the nuclear industry, are

considered to be acceptable. More information on the effect of low doses of radiation will be available within the next ten years as the result of a study which records the causes of death and the total lifetime dose of people who work in the nuclear industry. Once the risks are known accurately, then it is a matter of deciding the relative risks and benefits of artificially produced radiation and what risk we, as a society, think acceptable. In the meantime, the radiation protection ethos is the pursuit of ALARA — As Low As Reasonably Achievable. The problem is defining what is meant by 'reasonably', and of assessing not only the risks of *normal* levels of artificially produced radiation, but also the radiation levels that may result from accidents. In 1984 several beaches in Cumbria were closed because of contamination by radioactive plutonium and caesium, released into the sea by the nuclear power plant at Sellafield. In addition, there is concern about an increased incidence of deaths from leukaemia in local residents which may have resulted from environmental contamination.

Noise

Sound is produced and transmitted by the vibration of molecules. It travels through the air as a series of pressure waves. Humans 'hear' sound because these waves enter the ear and cause the eardrum to vibrate. These vibrations are transmitted by the bones of the middle ear and, in turn, cause very small vibrations in the hearing apparatus in the inner ear, which transmits electrical impulses via nerves to the brain.

The intensity of sound is measured in decibels (dB). Conversation occurs at about 50 dB; heavy traffic on a nearby road gives a noise level of 80 dB or so. 'Loud' music is usually at a level of 100 dB, the same level as a car horn about 0.5 metre away. Noise starts to be uncomfortable at levels of 120 dB and causes actual pain at 140 dB, which is the noise level close to a jet engine. At and above this level, noise can actually cause tears in the more delicate tissues of the body (particularly in the ear and brain) because the fluctuations in pressure that form the sound wave are large. More moderate, but continuous, levels of noise also cause damage to the ear, because the continual vibrations diminish the sensitivity of the hearing apparatus. Permanent deafness resulting from excessive background noise develops fairly rapidly (over a few years) and then gets worse more slowly with continued exposure.

The effects of noise vary with the intensity of the sound and with its duration. A background noise of 90 dB for forty hours a week over thirty years will cause significant deafness in 1 per cent of people. Hearing protection is considered advisable whenever there is a continuous background noise of 85 dB — that is about the noise level at which two people can just carry on a conversation if they stand close together and raise their voices. Hearing protection should be worn for any task where the noise level is more than 105 dB, even if it only lasts for a few minutes, such as at airports. Some concern has been raised about the levels of noise at pop concerts and discotheques, but so far there has been no national legislation in the UK limiting the noise level allowed.

Summary

In this chapter we have attempted to demonstrate some of the ways in which aspects of the physical environment affect human health. We have mostly used as examples *proximate* causal relationships, as they are the best recognised and documented: a lack of vitamin A and keratomalacia; freezing temperatures and frostbite; dioxin and chloracne; ionising radiation and leukaemia. However, as was indicated in the discussion on food, factors in the physical environment are also important *distal* causes of disease: for example, the association between diet and the development of coronary heart disease. In some instances these distal relationships have already been recognised, but there are many aspects of the physical environment that at present we only suspect of influencing health. Thus, for many environmental factors, we can only speculate about their impact.

Meanwhile, humans continue to pollute the atmosphere with motor and industrial emissions, nuclear radiations and noise; pollute the land with industrial wastes and pesticides; pollute the streams, rivers, lakes and seas with sewage, industrial effluent and oil. They disturb the environment, and in return the environment disrupts their health. The importance of the physical environment for health has been recognised since earliest times. Around 400 BC the Hippocratic school wrote of the impact of the weather, water supply, food consumption and the state of the air on human health. Human health partly depends on the maintenance of a stable relationship with the physical environment. If external homeostasis breaks down, then so may internal balance, with a consequent cost to health. How humans will respond to this is a matter of widespread concern. The alternative responses have been indicated by the plant biologist, Eric Ashby, for man at least:

What we are experiencing is not a crisis: it is a climacteric. For the rest of man's history on earth, so far as one can fortell, he will have to live with problems of population, of resources, of pollution. And the seminal problem remains unsolved: Can man adapt himself to *anticipate* environmental constraints? Or will he (like other animal societies) adapt himself only in *response* to the constraints after they have begun to hurt? (Ashby, 1978, p.3)

Objectives for Chapter 13

When you have studied this chapter, you should be able to:

13.1 Describe, with examples, how dietary deficiencies and obesity may affect health.

13.2 Explain how the two ways in which the body may respond to trauma — cellular repair and scar formation — depend on the nature and extent of the inflicted damage.

13.3 Describe, with examples, the four main ways in which chemicals may cause disease: (i) by damaging the structure of cells; (ii) by interfering with cell functions, such as enzyme systems and neurotransmitters; (iii) by damaging DNA in the cell nuclei; and (iv) by stimulating an adverse immune response (allergy).

13.4 Describe some of the adverse effects of different forms of radiation (including noise) and in particular the damage caused by ionising radiation.

Questions for Chapter 13

1 (*Objective 13.1*) What are the main threats to health posed by the diet consumed by people in the UK?

2 (*Objective 13.2*) Someone breaks their arm in a road accident. The injury involves both a fractured bone and severely torn muscles. How will the body repair the damage to these two tissues and which injury is likely to result in the greater disability in the long term?

3 (*Objective 13.3*) Cigarette smoking introduces tars, oils and carbon monoxide into the body. Describe three ways in which smoking may harm the normal functioning of cells.

4 (*Objective 13.4*) What is the most likely inanimate environmental cause of deafness in (a) a child living in a village in the Andes, (b) a boilermaker?

14
Living with other species

The discussion of infectious diseases and infestations in this chapter draws on the material on single-celled organisms in Chapter 3. During this chapter you will be asked to read an article in the Course Reader by Marc Strassburg (Part 4, Section 4.5). In addition, the article by Cecil Helman (Part 1, Section 1.2) is referred to.

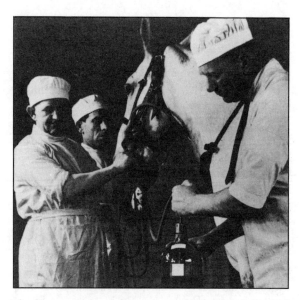

Blood being withdrawn from a horse which as been immunised against diphtheria. Anti-diphtheria serum was prepared from the blood and used to treat people with the disease (1944).

In the course of their lives humans interact not only with the physical environment but also with numerous plants and animals, large and small, as well as with other humans. Some of these other species are useful to humans, some are harmful, and some do not affect them at all. It is obvious that lions, poisonous spiders, stinging nettles and deadly nightshade can all cause harm in different ways to humans. In this chapter, however, we shall consider not these large, clearly visible hazards to health, but the much smaller organisms that cause *infections* and *infestations*. Just as the vast majority of large life-forms are harmless, so only a small proportion of small organisms can cause diseases. We are concerned with six groups of organisms: viruses, bacteria, fungi, protozoa, helminths and insects.

☐   What are the main differences between viruses and bacteria?

■   Viruses are composed of a chain of nucleic acid surrounded by a protein coat. They cannot reproduce on their own, but must enter the cell of a plant or animal and make use of that cell's synthetic machinery to replicate their own nucleic acid and synthesise the proteins for their coat. Bacteria are true cells, in that they are surrounded by a cell membrane. However, there is no nucleus — the bacterial DNA exists as a single strand in the cytoplasm.

Different species of bacteria have different needs, produce different enzymes and require different environments. As you may recall from Chapter 3, they come in three main shapes. They may be spherical (when they are known as cocci), rod-shaped (when they are known as bacilli), or spiral-shaped (spirochaetes). Each species of bacterium is known, like other organisms, by first a family name and then a species name (e.g. *Streptococcus pneumoniae*). Within a species there are variations between different bacteria, just as there are variations between different

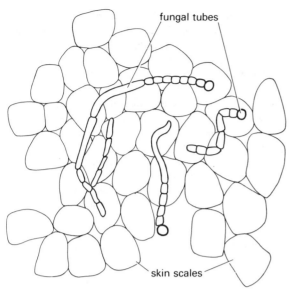

Figure 14.1 Athlete's foot (*Tinea pedis*) showing skin scales and the branching tubes of the fungus.

stage they may be visible to the naked eye, most have one stage as eggs or cysts which is microscopic. In general, helminths are not a major cause of disease in rich,

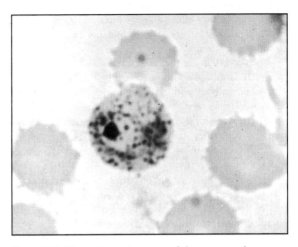

Figure 14.2 *Plasmodium vivax*, one of the protozoa that causes malaria, inside a human red blood cell (magnification 800 times).

humans. A group of bacteria within a species that show similar characteristics are said to belong to the same strain. In this chapter we shall use the names of several types of bacteria. Note that it is much more important that you remember how they behave and what effects they have, rather than their names.

The next group of organisms we shall consider are the fungi.

☐ What diseases do you know that are caused by fungi?

■ You might have mentioned athlete's foot (Figure 14.1), ringworm and thrush, a yeast infection (yeasts are small, round, single-celled fungi).

These are all common in the UK. Fungi do not only cause minor inconveniences, they also cause rare forms of pneumonia, which may be life-threatening.

The fourth class of organisms that can cause disease are protozoa. The most widespread diseases caused by protozoa are malaria (Figure 14.2) and some forms of dysentery.

☐ What are the characteristics of protozoa?

■ They are single-celled organisms, larger than bacteria. Some of them are just visible to the naked eye. They are more mobile than bacteria and many have the ability to change shape.

Next we come to the multicellular organisms, which strictly speaking do not cause infections, but infestations. The first of these are the helminths — the roundworms, flukes and tapeworms (Figure 14.3). Although in the adult

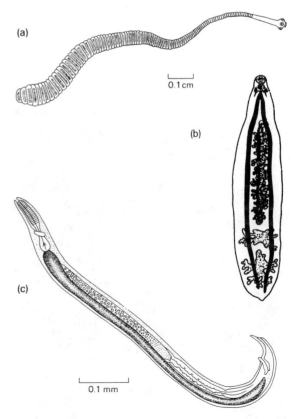

Figure 14.3 Helminths: (a) a tapeworm (*Taenia solium*); (b) a liver-fluke (*Clonorchis*); and (c) a roundworm (*Strongyloides stercoralis*).

industrialised countries with good sanitary systems, but they are extremely important in Third World countries. There they cause chronic debilitating diseases characterised by general lethargy and lowered immunity. Like protozoa, the helminths that infect humans often have complicated life cycles, living for one stage in humans and for another in some other animal.

The last group are the insects, ticks and lice. Although many are large and could not possibly inhabit the surface of the human body without being noticed, others, such as fleas, lice, ticks and mites are small and would pass unnoticed were it not for their effects. Figure 14.4 shows two that commonly live on humans.

The vast majority of organisms belonging to the groups mentioned above (with the exception of viruses) live independent lives feeding on dead organic matter — rotting vegetation or dead animals. Any small sample of garden soil will be found to contain huge numbers of bacteria, fungal spores, protozoa, roundworms and insects. In fact, it is largely because of the activity of these organisms that dead vegetation decomposes into soil. A small proportion of organisms (roundworms, flukes, and viruses) live and reproduce on or in other living plants or animals. They are known as *parasites* and the organism on or in which they live is, as you may know, called a host. The parasite does not necessarily harm the host. Parasites that simply live on or in their hosts are known as *commensals*.

If both the parasite and the host benefit from the relationship, it is known as symbiotic. Sometimes, when infectious diseases are being discussed, the word parasite is used to mean only the protozoa and helminths. Indeed, the people who study diseases caused by these organisms are known as parasitologists, whereas those who study bacterial and viral infections are known as microbiologists. We shall not be making such a distinction here and shall deal with diseases caused by both groups of organisms.

Somewhat paradoxically in a chapter on infectious diseases, we shall start by considering those organisms which live in or on human bodies but do *not* give rise to disease — the commensals. This has been done deliberately to emphasise the point that for most of the time humans are not impeded by their relationship with all the many different organisms with which they come into contact, and that disease is fortunately a rare outcome from such contacts.

Human commensals

Humans are host to large numbers of commensal organisms — on the skin, in the nose, mouth and throat, and in the small and large intestines. Each species colonises an area where the conditions suit it best. For example, the bacteria around the teeth and gums are different from the bacteria that live on the tongue, and these in turn are

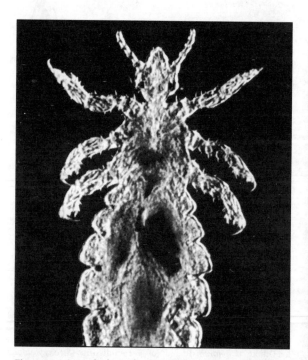 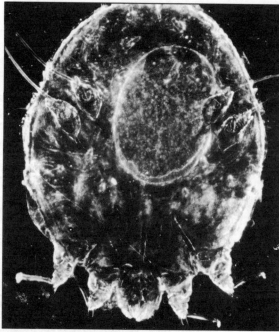

Figure 14.4 Animals that infest human skin: (left) head louse (magnification 30 times); and (right) scabies (magnification 75 times).

different from those on the mucous membrane lining the mouth, or those at the back of the nose. On the skin are types of bacteria that thrive in a slightly cooler, drier environment. They may co-exist with lice and fleas which are considered normal commensals in some cultures, though not in ours at present.

Conditions inside the gut are very different — warm, wet and with low levels of oxygen. Enormous numbers of bacteria grow there (10 000 million in each gram of faeces in the large intestine) and there may also be protozoa and helminths. Because of the low oxygen levels, a large proportion of the bacteria are anaerobic: that is, they grow better in low oxygen concentrations than high ones. The bacterial population is not necessarily stable: there are not always the same bacteria in the same numbers. If local conditions change, then the new environment may suit some species better than others. As a result, these will multiply faster and will come to make up a greater proportion of the population than they did before.

☐ From what you have just read, would you expect the bacteria in the gut of a breast-fed baby to be the same as those of a baby fed on cows' milk?

■ No. The chemical composition of human breast milk and cows' milk is different. This means that the environment in the gut will be different.

Human breast milk contains a high proportion of lactose (a type of sugar). This favours the growth of bacteria which feed on lactose and produce acetic acid (vinegar). This keeps the contents of the gut acidic. This acid environment suits certain types of bacteria which are not found in the gut of bottle-fed babies, where the environment is more alkaline. This accounts for the different appearances of the faeces of infants who are breast-fed from those who are bottle-fed.

The vagina is another area where the degree of acidity varies. The amount of glycogen, a carbohydrate that is stored in epithelial cells lining the vagina, varies under hormonal influence. The normal inhabitants of the vagina vary as the glycogen content changes during the menstrual cycle. High glycogen levels favour acid-producing bacteria.

Not all of the body is colonised by commensal microorganisms. The bones, muscles, blood and other tissues that lie between the skin and the gut are normally sterile (free from bacteria or other microorganisms), as are the brain and spinal cord, the kidneys, bladder and urine, and the joints. Figure 14.5 shows the so-called 'microbiologically dirty' areas of the body.

In general, commensal microorganisms cannot get through the skin or mucous membrane on which they live. They stay on the surface or perhaps live in the opening of a gland. There are, however, sometimes small wounds or breaks in this mechanical barrier and the microorganisms

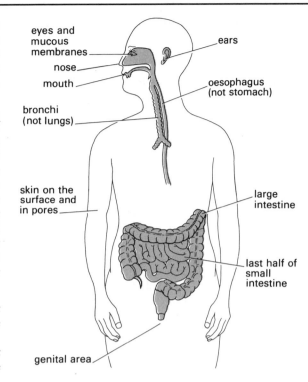

Figure 14.5 Microbiologically 'dirty' areas of the human body.

can then pass into the deeper tissues.

☐ Can you recall what happens to them there?

■ They are recognised as 'non-self' and engulfed by macrophages. If there are enough of them, they will provoke first an inflammatory and then an immune response. Under normal conditions, therefore, microorganisms do not colonise areas where there are scavenging white blood cells.

In areas where colonisation normally occurs, the populations of microorganisms are fairly stable, as long as environmental conditions are steady, and they do no harm. Competition for food and oxygen between different bacteria prevents their multiplying to excess.

When do organisms cause disease?

We have taken some time here to stress that humans co-exist with many microorganisms without problem. On the other hand, this chapter would not be here if microorganisms did not sometimes cause disease. Indeed, 100 years ago in the UK, half of all deaths were from infectious diseases and, in Third World countries, this is still true today.

☐ From what you have read so far, what factors determine whether someone gets an infection?

■ (i) Environmental conditions. If these change, then

they may favour the reproduction of one particular sort of microorganism rather than another. (ii) The integrity of physical barriers like skin and mucous membranes. (iii) The competence of the person's immune system (that is, their resistance to infection). (iv) The virulence of the microorganism (discussed in Chapter 11).

Some organisms have developed methods of getting into the tissues of the body. The malaria parasite, for instance, is injected into the bloodstream by the mosquito in whose salivary glands it spends part of its life cycle. There are bacteria that have features which help them to spread through mechanical barriers within the host tissue. For example, some *Streptococci* produce an enzyme which catalyses the breakdown of substances between the host's cells, so that the cells separate more easily. For this reason, *Streptococci* often produce spreading infections (cellulitis) rather than localised ones (such as an abscess or boil).

Some organisms are not easily destroyed by the primary defence mechanisms, the inflammatory and immune responses. Strains of *Streptococci pneumoniae* can, as the name suggests, cause pneumonia (inflammation of the alveoli in the lungs). These particular bacteria are virulent because they have a carbohydrate capsule surrounding their cell wall. This capsule prevents their destruction by macrophages. Other strains of *Streptococci pneumoniae* do not have this capsule and as a consequence rarely cause disease. Tubercle bacilli that cause tuberculosis have a waxy coat. They can be engulfed by macrophages but not digested. In fact, they can multiply inside the macrophage, eventually killing it so that they are released into the tissues again. Yet another protective mechanism is shown by *Staphylococci* which produce an enzyme which coagulates blood plasma. The plasma forms a protective coat around the bacteria which prevents the macrophages from engulfing them.

All these mechanisms provide means for microorganisms to survive their first few days in the host tissues. Ultimately, how far they spread depends on the immune response of their host. This is not necessarily the same in all people, nor is it constant throughout life in one person. One consequence of such differences in host response is that the same microorganism can cause different levels of infection in different people. For example, *Meningococci* are frequently found living in people's throats without causing any signs of illness. Yet, in other people, these same bacteria can cause meningitis — inflammation of the layers (or meninges) covering the brain and spinal cord.

Some organisms do not cause a large immune response. In some cases, this is because their surface antigens are very similar to some host molecules. Figure 14.6 shows two sorts of ringworm in humans. One is caused by a fungus that usually infects humans and the second, by a similar fungus whose usual host is cattle.

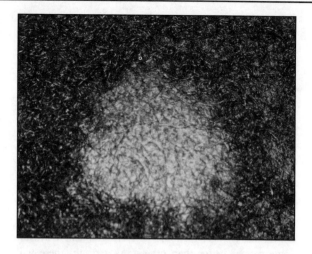

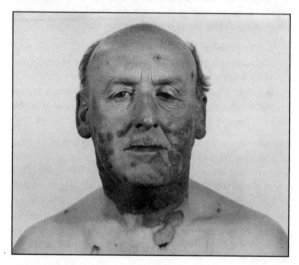

Figure 14.6 Ringworm (a) bald patch on scalp caused by *Tinea capitis*, and (b) patches on the face and neck caused by cattle ringworm.

☐ Which organism do you think would provoke the greater immune response in humans?

■ The cattle ringworm fungus. This is because its antigens are much more dissimilar to human molecules than the antigens of human ringworm.

In other cases, such as roundworms that live in the human gut, a poor immune response results from the thick waxy skins (cuticles) they possess. Although the cuticles are made of protein, they do not seem to stimulate a response in their host, perhaps because they are covered in mucus.

The influenza virus shows a characteristic that enables it to infect the same person more than once. It stimulates, in the normal way, the production of antibodies to the antigens in the protein coat of the virus. These antibodies continue to circulate for many years, so theoretically humans should get 'flu only once in a lifetime. However, the antigens of the protein coat of the influenza virus are not constant and they undergo a major mutation, a so-called shift, about every ten years. This means that people are no longer immune to the 'new look' virus and can again become ill with influenza. This is why there are major epidemics of 'flu about every ten years.

Other organisms actually change their antigenic coat during their life cycle. Antibodies produced by their host against their early forms no longer affect them as they mature. Schistosomes, flukes that cause bilharzia, show this effect, as do *Plasmodia*, the protozoa that cause malaria. The best example, however, is given by the protozoan (*Trypanosoma*) that causes sleeping sickness in Africa. These protozoa change their surface antigens frequently — there are at least twenty-five variants. The organism is, therefore, protected against the host's antibody production, which is always 'one step behind' the trypanosome's surface antigens. Schistosomes also avoid being damaged by the host's immune system by disguising themselves in a covering of host molecules. Thus the host fails to detect them.

How do organisms cause disease?

We have now established how organisms can get into people and start to multiply, but this is not always immediately obvious to the infected person. Usually there is an *incubation period* during which there is little sign of anything happening. This may last from a few hours (bacterial dysentery) to many years (some so-called slow virus diseases). Most commonly, though, the effects start to show after a few days, perhaps with a fever, aches and pains, a rash or a runny nose. What causes these signs and symptoms by which an infection is recognised? As you might imagine, there are a variety of causes.

Some symptoms are due entirely to chemicals, produced by the infecting organism, known as *toxins*. They may be secreted by the live organisms, when they are known as exotoxins, or they may be intracellular substances released when it dies inside the host, called endotoxins. Tetanus is one example of an infection whose symptoms are caused entirely by an exotoxin. The bacteria causing tetanus are found in soil and dirt and may get into skin wounds, where they multiply and produce tetanus toxin. This toxin is taken up by nerve endings and it travels up the nerve to the spinal cord and brain, where it disrupts the control of muscular activity. The final result is intense muscle spasm, the symptom that gave tetanus its common name of lock-jaw. The toxin causes widespread effects, even though the bacteria are still localised to the skin wound.

Endotoxins have many different effects. Some cause the small blood vessels to both dilate and leak more fluid into the surrounding tissues. If the infection is widespread throughout the bloodstream (known as septicaemia), dilation of all the blood vessels in the body will occur.

☐ What effect do you think this will have on the heart rate and blood pressure? (You may wish to refer back to Chapter 8.)

■ Dilation of the blood vessels and increased leaking of fluid into the tissues will cause a fall in blood pressure. The heart rate will increase in an attempt to restore the blood pressure. If most of the blood vessels are dilated, this response will be inadequate and the blood pressure will continue to fall. A catastrophic fall in blood pressure is often the cause of death in septicaemia.

Endotoxins are also responsible for causing fever, although there is evidence that pyrogens ('fever producing' substances) are also released by macrophages and other damaged host cells. Apart from those symptoms caused by toxins, most of the symptoms and signs of illness are due to the host's inflammatory and immune responses to non-self molecules.

☐ Why do you think your nose gets blocked up when cold viruses invade the cells at the back of the nose? (You may wish to refer back to Chapter 9 on the acute inflammatory response.)

■ The blood vessels in the mucous membrane of the nose dilate and become leaky. Fluid passes out from the capillaries into the surrounding tissue. The mucous membranes thus become swollen and 'block' the back of the nose. In addition, some of the fluid leaks out of the damaged surface of the mucous membrane, causing the nose to 'run'.

All this is due to the acute inflammatory response. In some diseases it is the later antibody response that actually causes the symptoms. The rash in measles (a virus infection) appears when the body starts to produce antibodies and it is thought to be caused by antigen–antibody complexes causing localised inflammation in the skin. This accounts for the puzzling observation that people with lowered immunity become very ill when they get measles and yet they do not develop a rash.

Sometimes antigen–antibody complexes can cause damage after the original infection is over. A streptococcal throat infection usually lasts about a week and then clears up. In some cases, however, it is followed, about three weeks later, by inflammation of the kidney. This kidney damage, which may be permanent, is caused by the deposition of such complexes in the kidney.

Toxins and the host's inflammatory and immune responses are the commonest causes of symptoms, but there are other factors. Occasionally there may be competition for resources (for example, for a chemical) between host and parasite. The hookworm (a roundworm found in tropical areas) has a large mouth with sharp teeth (Figure 14.7). It attaches itself to the wall of the small intestine with its teeth and feeds on blood from the lining of the gut. The blood loss that results makes hookworm infestation the commonest cause of anaemia in many parts of the Third World. Tapeworms also live in the gut, where there is usually enough food for both the worm and its host. However, one type of tapeworm consumes such large amounts of vitamin B12 that it may cause a deficiency in the host.

Occasionally parasitic organisms cause mechanical damage. The eggs of *Schistosoma* have a sharp spine which cuts through tissue. The commonest roundworm, *Ascaris*, is often found in large numbers in the gut. There may be up to 100 in number, entwined to form a round mass which can completely block the gut.

Some infections can pass from a mother to her fetus, and a few of them may cause damage to the developing fetus. The best known of these is rubella, or German measles. If a woman acquires the infection for the first time during a pregnancy, the virus may pass from her bloodstream into the cells of the uterus that are just next to the cells of the placenta. Damage to these cells means that the virus is able to pass into the placenta and, hence, to the fetus itself. Here it affects the rate of mitosis of cells. Newborn babies who have had rubella during the first few months of pregnancy have fewer cells in each organ than normal babies. Rubella can cause major deformities in any organ, but the commonest results are malformations of the heart, blindness, deafness and mental retardation. The effects are more severe the earlier in pregnancy the infection occurs and infection during the last four months of pregnancy is usually harmless.

□ Why do you think this is so?

■ Because the main organs *develop* during the first three months of fetal life. After this, mainly growth, rather than development takes place.

Another virus that may cause congenital abnormalities, if the mother gets it for the first time during pregnancy, is cytomegalovirus. This has been estimated to occur in 1–2 per cent of all pregnancies, but only occasionally leads to damage to the fetus. Cytomegalovirus usually causes a very mild illness in the adult, but the fetal response varies from no apparent effects to a severe illness, including jaundice and platelet disorders, and, if the baby survives, deafness, blindness, problems of bone formation and mental retardation. It has been estimated that cytomegalovirus could be responsible for brain damage in several hundred babies a year in the UK.

Now that you know how organisms cause illness, you are in a position to think about how infections might be prevented and treated.

□ From what you read in the previous section, what two methods could you use to prevent an infectious disease from occurring?

■ You could try to stop the organism getting into a potential host, or you could try to increase the host's response to any organisms that do get in.

The third response is that of treatment, the use of drugs to kill off infective organisms within the body. For the rest of this chapter we shall be considering these three subjects: the prevention of the spread of organisms, raising the infective host's immunity, and treatment.

Prevention of spread

To prevent the spread of infective organisms, it is first necessary to know how they are transmitted. If you get measles, for instance, where did the measles virus come from and how did it get into your body? There are five main sources of infection: the organism may already be present as a commensal, or it may be transmitted by air, by physical contact with an infected person, by food and water or by an insect.

You have already read about the conditions under which commensals, normally benign microorganisms, may cause illness.

□ Can you remember what these are?

■ (i) Loss of mechanical barriers; (ii) change in environmental conditions so that one type of organism can multiply to excess; (iii) a lowering of the host's ability to respond to non-self molecules.

The general immune responsiveness may be lowered under several circumstances: it is sometimes low in old age; it may

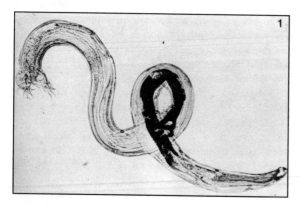

Figure 14.7 Hookworm. Note the mouthparts and the (human) blood in the worm's digestive tract.

be decreased after certain viral infections, particularly after measles, influenza and glandular fever; and it may be lowered artificially by means of drugs, such as those given to people undergoing kidney transplantation. People whose immune system is suppressed, for whatever reason, are at risk of falling ill from normally non-pathogenic organisms.

A change in the environment may be caused in several ways. As was mentioned earlier, the acidity of the vagina varies during the menstrual cycle. During pregnancy, or under the influence of the contraceptive pill, the vaginal contents become more acid. This acidity favours the growth of some types of fungi (yeast) which are present as normal commensals in the vagina of 25 per cent of women. If the fungi multiply beyond a certain number, they change their behaviour. Instead of being round, unicellular organisms, they grow into thin branching tubes which grow into the mucous membranes of the vaginal wall. This causes intense itching and a white vaginal discharge, a condition known as thrush or *Candidiasis*. *Candidiasis* is also common in women who have been receiving antibiotics for an infection elsewhere in their body.

□   Can you think why this might be? (Remember the discussion earlier on the balance between commensals.)
■   Antibiotics will destroy some of the commensal bacteria in the vagina which means that more nutrients are available for other commensals, such as fungi, that are unaffected by antibiotics. The fungi are free to multiply and consume the extra food that is available.

Vaginal thrush is an example of the excessive multiplication of a commensal in its usual home. Our next example is of an infection caused by a commensal in an area of the body that is usually sterile. Cystitis is the name for an infection of the bladder wall and the urethra, the main symptom being pain on urinating.

□   The bacteria that cause cystitis are usually commensal bacteria in the 'wrong place'. Where do you think they come from?
■   Usually from the gut or external genital area. The external openings of the urethra and large intestine (anus) are close together. This means that normal gut commensals can sometimes get into the urethra and pass up into the bladder.

Cystitis is more common in women than men. This is thought to be because the female urethra is shorter: the bacteria do not have so far to travel to the bladder, where they can breed. Most of these bacteria prefer an acid environment. Hence, the traditional treatment for cystitis is to drink potassium citrate solution. This makes the urine less acidic and kills off the bacteria, or at least makes them less likely to multiply. Another traditional remedy is simply to drink a lot — the more the bladder is emptied, the more often bacteria are washed out with the urine.

'Coughs and sneezes spread diseases' goes the saying, and it is certainly true for many of the more common infections — for viral infections such as mumps, measles, rubella, chicken-pox, colds and influenza, and for some bacterial infections such as whooping cough, diptheria, tuberculosis and many other respiratory infections. Colds are the most common example. They are usually caused by rhinoviruses ('rhino' meaning nose) which prefer a temperature of about 33 °C as found at the entrance to the nose. Rhinoviruses usually cause quite a mild illness of a runny nose for three to four days. When someone with a cold coughs or sneezes, they give out small droplets of fluid that contain the cold virus. Other people nearby breathe in these virus particles and some of them develop the infection. Despite producing antibodies to the cold virus, which continue to circulate for up to a few years, people may get several colds each year.

□   Why do you think people can repeatedly get colds?
■   There is not (unfortunately) only one type of cold virus. There are more than 100 different rhinoviruses, each with its own surface antigens.

This may explain why adults get colds less often than children. Each time someone has a cold they develop some immunity to that particular rhinovirus and during a lifetime will gradually become immune to most rhinoviruses.

Tuberculosis (TB) is a bacterial infection that, in one form, may be spread through the air from an infected person. The tubercle bacilli are breathed into the lungs where they lodge and slowly produce cavities filled with cheesy-looking pus. The contents of these cavities may be coughed up together with tubercle bacilli which may in turn infect other people. The notices in public places, from Victorian times onwards, enjoining people not to spit, aimed to reduce the transmission of TB. You may remember that the tubercle bacillus has a waxy coat. This enables it to survive in dry conditions, so that bacilli can lie dormant in dust for years until they reach an environment suitable for multiplication. Most other bacteria and viruses can exist for only a short time in the air or dust, since drying out kills them.

□   How would you try to control an illness caused by a virus that is mainly spread via the air?
■   The short answer is that it is very difficult. Isolation of infected people is possible, but most people are infectious (that is, give out the virus) for about a day before the symptoms appear. Good ventilation in crowded places may help, since the virus particles are then diluted quickly and no one person receives a large dose. Preventing the spread of airborne infections is, however, not usually a viable means of control and many airborne infections are still endemic even in countries that have good public health measures.

You may wonder why TB is not still common in the UK, if the bacillus has the extra advantage over viruses and other bacteria of surviving for long periods outside the body. The answer seems to lie in the fact that the incidence of TB is related to the standard of living. As people become better fed and better housed, they seem to become less susceptible to infection by the TB bacillus.

Another category of diseases are those spread by skin contact. Most of these are minor infections of the skin, though some of them, such as leprosy, cause serious disease and have their major effects in other parts of the body. Among the minor infections are those caused by *Staphylococci* and by viruses (warts and verrucae). The various lice and nits that live on the skin or hair of the human body may also be acquired by direct contact with an infested person.

Leprosy is of particular interest because it causes two main types of disease depending on the immune response of the sufferer. In people with a poor immune response, the leprosy bacilli multiply throughout the body and especially in the skin where they trigger a chronic inflammatory response. The skin becomes swollen and thickened and may produce exaggerated folds in the face. People with this sort of leprosy (lepromatous leprosy) remain infectious, though less so than popular myths would suggest (Figure 14.8). People with a high resistance to the leprosy bacillus show quite different evidence of the infection. The bacilli are confined to the nerves and to isolated patches of skin. The bacilli stimulate the production of lymphocytes, which may form a dense mass in the areas of infection. It is this swelling caused by lymphocytes that compresses the affected nerves and causes the numbness of the hands and feet and muscle weakness that is characteristic of this type of leprosy. People with tuberculous leprosy, as it is called, are rarely infectious.

One special group of diseases acquired by direct contact are those transmitted during sexual intercourse. They were previously called venereal diseases, but are now referred to as sexually transmitted diseases. In the UK, these are the only diseases spread by skin contact that arouse much interest. The mainstays of control are treatment of infected people and contact-tracing, organised via 'special clinics'. There is considerable publicity to encourage people who think they may have a sexually transmitted disease to attend such clinics for a check-up. One of the difficulties in eliminating a disease such as gonorrhoea is that it may pass unnoticed, especially in the female, and it is always difficult to control the spread of any disease when the microorganism responsible may be carried without any symptoms. Though gonorrhoea *may* be symptomless in men, it generally causes extreme pain on urinating and a discharge from the urethra. In women, it often causes no symptoms beyond a slight vaginal discharge. The woman is then a

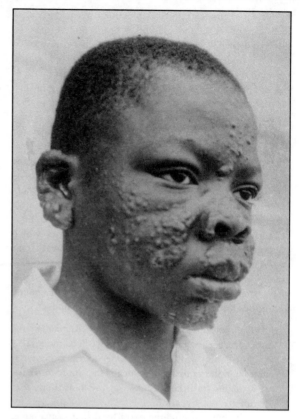

Figure 14.8 Lepromatous leprosy affecting the face.

carrier and unknowingly can pass on the infection. Table 14.1 lists the common sexually transmitted diseases.

Recently, genital herpes has gained much publicity. It is caused by the *Herpes simplex* virus that can also cause cold sores around the mouth, although the latter are more often due to a slightly different strain of the virus. The *Herpes simplex* virus is an example of a virus that can remain latent ('resting') in cells. Most people are infected by the virus as children: they may have a low fever and some ulcers in the mouth or sores on the face, or suffer no symptoms at all. They produce antibodies against the virus, but the virus is not completely eliminated from the body — it remains latent in nerve tissue. Under certain conditions, such as exposure to sunlight, or during some other viral infection, or at a time of emotional stress, the *Herpes* virus becomes active and causes cold sores on the face. They heal after a few days and the virus returns to a latent phase in the nerve tissue. The same process occurs with the genital version. Victims of genital herpes will suffer acute episodes of illness separated by latent periods when they are free of any symptoms. However, even during such periods, it appears that sufferers of genital herpes may be infectious and can transmit the virus to sexual partners.

Table 14.1 The common sexually transmitted diseases

Disease	Type of microorganism	Symptoms
syphilis	bacterium	initially, ulcers on the skin or mucous membranes at site of entry to the body; later, widespread effects in the body affecting the nervous system
gonorrhoea	bacterium	men — pain on urinating, discharge from urethra; women — usually symptomless
non-specific urethritis (NSU)	sometimes chlamydia (a type of bacteria)	men — pain on urinating, discharge from urethra; women — usually symptomless
genital herpes	virus	recurrent small painful ulcers in genital area
trichomonas	protozoon	men — irritation around urethral opening; women — vaginal discharge, itching
candidiasis (also known as thrush)	fungus	similar to trichomonas
genital warts	virus	soft warts in genital area
pubic lice (also known as 'crabs')	insect	itching and bites
cervical cancer?	virus?	none initially; bloodstained vaginal discharge, and pain in advanced cases

Some diseases are spread via food and water. Many microorganisms that live in the gut are excreted in the faeces, as are the eggs of worms living in the gut. Obviously, if they are to breed and reproduce successfully, they must get back to another human host. This happens in several ways. Some pathogens are transmitted back to their hosts via water, either in drinking water or through the skin. John Snow, in the mid-nineteenth century, showed that the cholera epidemics in London were due to contamination of the water supply by sewage. The bacteria that cause cholera, ingested in drinking water, multiply very fast in the small intestine. They produce two exotoxins that effect transport mechanisms in the cell membranes in the wall of the small intestine: fluid is not absorbed from the gut into the body, but passes from the body into the gut, and is lost as watery diarrhoea. People with cholera lose so much fluid in diarrhoaea that dehydration is the major cause of death. The diarrhoea is then full of the cholera bacteria. For the disease to spread, the bacteria must get from sewage into the water supply. If standards of sanitation are poor, one cholera victim can infect a whole community. Poor sanitation is one reason why cholera epidemics occur so frequently in refugee camps. The disease is found only in humans, so water supplies cannot be contaminated by animals.

☐ What do you suppose is the best way to prevent cholera epidemics?

■ The strict separation of sewage disposal and drinking water supplies.

At present, only about one-third of the world's population has water supplies that are safe in this way.

Typhoid is another bacterial infection that is usually spread in drinking water, though sometimes in contaminated foods. It differs from cholera in that the bacteria are often carried by people in their gut without causing illness. Carriers of typhoid excrete the organism in their faeces and therefore may unknowingly be the source of an epidemic of the disease if there is some route whereby the bacilli can reach drinking water or food. In one outbreak of typhoid in Scotland in 1964, the source of infection was found to be tins of corned beef from Argentina. The source of *their* contamination was the river water that was used to cool the tins during the canning process. Although the river had always been contaminated, the water had been inadequately chlorinated on that occasion.

In the Third World, two important helminths (hookworm and *Schistosoma* fluke) get into their host through the skin — via soil or stagnant water in puddles and canals. The adult hookworm lives in the small intestine, where it causes anaemia by feeding on human blood. It lays enormous numbers of eggs, up to 30 000 a day, which pass out in the faeces and hatch in wet soil as small wormlike larvae. These are free-living for a while, but gradually develop into infective forms that can penetrate the skin of someone walking in the damp soil. An itchy spot occurs where they enter. The larvae pass into the bloodstream and then into the lungs. They are then coughed up, swallowed, and finally arrive at their adult home, the small intestine.

☐ What three conditions are necessary for the hookworm to grow from egg to adult?

■ (i) Water or wet land; (ii) people who walk barefoot; (iii) people who defaecate in open places.

Somewhat similar conditions also suit the *Schistosoma* fluke that causes bilharzia. There are three different types of *Schistosoma*, though their life cycles are essentially similar. Two types live their adult life in the veins around the gut: the third lives in the veins around the bladder. The adults lay eggs which have a sharp spine (Figure 14.9). The egg tears through the gut or bladder wall, and passes out of the body in the faeces or urine. The eggs hatch in water and the larvae infect their intermediate host, the aquatic snails that live in canals and irrigation ditches. From the snails the more mature organism returns to the water for a time before infecting more humans, again through the skin.

Schistosomes do not cause acute disease; the main damage to the host is caused by the blood loss and inflammation that occurs when the eggs tear the bladder or intestinal lining. Bilharzia, however, is a debilitating disease that lowers vitality, strength and immune status, so that affected people are more susceptible to other infections. There is also evidence that cancer of the bladder is more common in people with bladder schistosomiasis.

☐　How would you suggest trying to prevent the spread of schistosomiasis?

■　You could try killing the intermediate host (snails) by chemical means; encouraging people to wear protective clothing when working in wet fields; stopping children from bathing in irrigation canals; and installing proper sanitation so that sewage does not get into the irrigation canals.

This all sounds so efficient that you may wonder why bilharzia is such a widespread disease. In fact, much effort has been expended on trying to control the snail population, but with little success. Wearing protective clothing sounds fine until you imagine wearing wellington boots all day in a hot damp environment. Good sanitation, which in the Third World normally means the provision of pit latrines, would gradually solve the problem, but such cabins become foul-smelling in the heat and represent a major change in people's sanitary habits.

Some infections are spread from faeces via food. They include dysentery — acute diarrhoea caused either by the bacillus *Shigella*, or the protozoan *Entamoeba* — and infectious hepatitis, a viral infection of the liver characterised by lethargy, fever and jaundice. It is a less serious infection than serum hepatitis, which is spread by contact with blood and is particularly found among drug addicts who share syringes.

Although most helminths are transmitted to humans by water or food, people become infested with tapeworms by eating the meat of the intermediate host. Figure 14.10 shows the life cycle of the beef tapeworm. A tapeworm living in the gut of a human drops off a segment of its body which is full of eggs. This is passed out with the faeces. It then breaks up, releasing the eggs. The eggs are ingested by beef cattle, the intermediate host, and eventually end up as

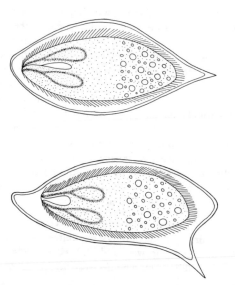

Figure 14.9 Eggs of the fluke which causes bilharzia, *Schistosoma mansoni.*

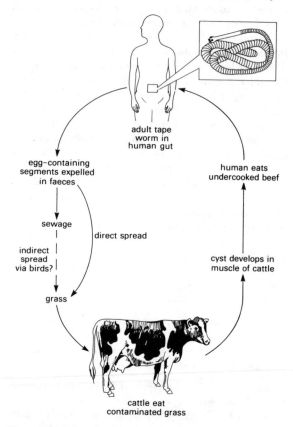

Figure 14.10 Life cycle of a tapeworm (*Taenia saginata*)

cysts in the muscle of the animal. If this is then eaten as inadequately cooked meat, each cyst has the potential to develop into a tapeworm in the human gut.

    ☐ How would you control the spread of tapeworms?
    ■ Good sewage disposal; inspection of slaughtered meat; ensuring that steaks are well-cooked.

In fact, improved standards of meat inspection at slaughter-houses has virtually eliminated the risk of infected meat being sold to the public in the UK.

Finally, there are some diseases in which the source of the organism is faeces, but which are transmitted by direct contact, rather than through contaminated food or water. These include polio virus, threadworms, whipworms and roundworm (*Ascaris*). The last two both lay eggs in the gut which pass out in the faeces. These eggs can withstand drying out and many disinfectants. They can live in dust, on hard surfaces (such as lavatory door handles) or on the hands, until chance brings them to someone's mouth. If the eggs are ingested, they develop into adult worms in the gut. Threadworms (or pin-worms as they are also called) have a particular way of making sure their eggs reach the hands of their host. The adult threadworm (which is about 1 cm long) creeps out of the anus at night and lays her eggs on the skin, where she then dies. Her movements cause intense itching of the area around the anus. People therefore scratch, the eggs get stuck under their fingernails and may, as a result, be carried to the mouth.

    ☐ Can you suggest why threadworms, whipworms and roundworms exist in the UK whereas hookworms do not?
    ■ The control of hookworm depends on public sanitation methods. The control of threadworms, whipworms and roundworms depend on personal hygiene. It needs only one infected child or adult to spread the eggs to other people.

In fact these worms do not cause much illness, unless they are present in large numbers when appropriate drug treatment can get rid of most of them.

For some microorganisms, blood-sucking insects provide the method of getting through the human skin. For example, plague is caused by *Yersinia pestis*, a bacterium which can infect the domestic rat and the fleas it carries. When a flea becomes infected, its stomach becomes full of proliferating plague bacilli, which means that no fluid can get through for absorption into its body. This apparently leaves it extremely hungry. When it leaves the corpse of its host rat and finds a new host, which may be a human, it bites the new host again and again in an effort to get food. Each time it does so, plague bacilli are regurgitated into the wound. The bacilli are carried, via the lymph vessels, to the lymph nodes of the new host. If an infected human is in good health and the dose of bacilli is not too large, the bacilli are restricted to the lymph nodes by an acute inflammatory reaction, which results in the formation of the painful black pus-filled swellings known as buboes. If the infected human has low resistance to the bacillus, it may spread into the bloodstream and pass into the lungs, causing pneumonic plague.

Typhus (a different disease from typhoid) and malaria were major insect-borne diseases in Europe at the same time as plague. Although these diseases are no longer common in Europe, in tropical countries they are still of great importance. Malaria, for instance, is by far the most important cause of death in many parts of the Third World. *Plasmodium*, the protozoan that causes malaria, is carried by mosquitos, which act as the intermediate host. There are many species of *Plasmodium*, but only four infect humans and produce four different malarial diseases of varying severity. *Plasmodia* are injected into the bloodstream from the salivary glands of mosquitos. They spend about half an hour in the blood before passing into the liver where they grow and multiply for 7–10 days, before being released once again into the blood. Here they enter red blood cells and multiply once again, eventually causing the red blood cells to rupture. Up till this time their host feels no ill effects, but when the red cells rupture, pyrogens are released into the blood. This explains the sudden onset of high fever during bouts of malaria. While the *Plasmodia* are free in the blood they infect any mosquitos that bite the person. This starts the next stage of their life in the insect. In three of the four types of malaria, *Plasmodia* may remain dormant in the liver and emerge at some later date. This explains why people who have once had malaria may have relapses many years later.

A World Health Organization programme to control malaria by controlling mosquitos had some initial success

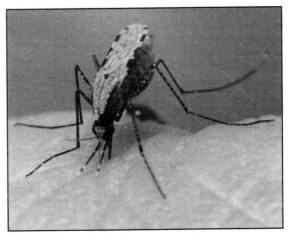

Figure 14.11 Female mosquito (*Anopheles gambiae*) in the act of feeding on human skin.

in India and Africa where epidemics of malaria were intermittent. However, malaria is once again on the increase, partly as a result of the mosquito population becoming resistant to DDT — the insecticide used.

Raising host immunity

We have mentioned several diseases where control of spread has not proved possible. Another method of controlling infectious diseases is the alteration of the host's immunity.

□ How is it possible to increase someone's immunity to a particular microorganism?

■ Lymphocytes are activated and antibodies are produced when non-self molecules get into the human body. If you can artificially introduce the antigens from a particular microorganism, then people will produce antibodies or lymphocytes against that microorganism.

This is the basis of *immunisation*, the artificial raising of immune response to a particular microorganism by the injection of a vaccine. This *response* will fade after the immunisation course, but the enhanced *ability* to respond will not. The heightened immunity is retained because the immune system adapts by increasing the number of lymphocytes capable of responding to that antigen.

For some infectious diseases, the production of a vaccine is a simple matter. If the illness is caused mainly by an exotoxin, as in diphtheria and tetanus, samples of the toxin can be modified so that, although it is no longer poisonous, it is still antigenic. It is then known as a toxoid. Someone into whom the toxoid is injected will then form antibodies against the toxoid. It is still possible to be infected by tetanus bacteria, but the exotoxin will be 'mopped up' by the antibodies and thus prevent the disease progressing.

For other diseases, such as typhoid, cholera, and whooping cough, it is difficult to define which antigens on the bacteria stimulate the antibodies that protect against infection. Therefore, vaccines are usually made from killed preparations of the organism. The protection is not usually as effective as it is against exotoxins, but nevertheless may prevent the disease in 60–70 per cent of immunised people. In addition, it makes the infection less severe in those who do get it. Yet other vaccines are made from live organisms that have been treated so as to reduce their virulence. The vaccines against measles and tuberculosis are of this type. It is obviously very important that live vaccines are carefully produced and that care is taken to ensure that the organisms really are non-virulent, so that they stimulate an immune response without actually causing the disease.

It has not proved possible to make effective vaccines against all infectious agents. There are two main reasons for this. The first is that for some diseases it is not possible

to get adequate supplies of the antigen because the bacteria or protozoa grow only in humans or rare animals. This is the main reason why there are, as yet, no effective vaccines against leprosy and malaria widely available. Leprosy bacilli grow only in humans and the nine-banded armadillo, and there are not enough armadillo to make vaccine manufacture commercially viable. New techniques of biotechnology, however, mean that antigen production is now both easier and cheaper. Vaccines for leprosy, rabies and syphilis are all in limited production, although clinical trials may take some time to establish their value.

The second factor that has impeded the development of some vaccines is exemplified by trypanosomiasis (sleeping sickness).

□ Why do you think it is difficult to make an effective vaccine against trypanosomes?

■ The surface antigens of trypanosomes change frequently, so that vaccines containing only a few antigens will quickly become 'out of date'.

Problems are also seen with schistosomes as a result of their ability to pick up human molecules, incorporate them in their own surface membrane and thus making themselves antigenically 'invisible' to their host. There is at present considerable research being done into the life cycle of these sorts of parasites which escape their host's immune response, in an effort to find one stage in their life cycle when a vaccine might prove effective.

Successful immunisation programmes depend on more than the preparation of a successful vaccine. Some programmes depend for their success on protecting not just individuals, but the whole community. If more than 70 per cent of people within a community are resistant to a disease, then there may be isolated cases of that disease but it will not spread easily and epidemics will not occur. Measles epidemics used to occur in the UK every two years as the number of children who had never had the disease rose towards one million. As a result of measles vaccine being given routinely to infants, epidemics no longer occur. Despite the vaccine not giving complete protection to each individual child, the effect of a large proportion of the childhood population being immune has proved sufficient to prevent the two-yearly epidemics, though insufficient to eradicate the disease.

Most vaccines carry some small risk of harming those to whom they are given. This risk to the individual must be balanced against the risk to individuals if the level of immunity within the community falls and epidemics occur. Pertussis (whooping cough) vaccine is thought to cause damage to the nervous system in about 1 in 300 000 recipients: that is, about two cases a year in the UK. In the whooping cough epidemic of 1977–79, which occurred when the immunisation rate among those aged under five had fallen to around 30 per cent, there were twelve deaths, mostly in children under the age of one. It is unlikely that the epidemic, and the deaths, would have occurred if immunisation rates had been maintained at above 70 per cent during the preceeding four years.

We shall end this discussion of immunisation with a success story — that of smallpox eradication. Smallpox was the first infectious disease for which a vaccine was found and the present-day vaccine gives very good protection against the disease. In 1966, the World Health Organisation (WHO) began a global campaign to eradicate the disease. A description of that campaign appears in the Course Reader in an article by Marc Strassburg (Part 4, Section 4.5). Read it now. When you have read the article, answer the following questions.

  ☐ What are the three important features of smallpox that were essential to the success of the campaign?
  ■ (i) Smallpox is caused by a very virulent virus that always causes disease in humans. There is no carrier state and, therefore, no human reservoir of infection. (ii) It affects only humans. There is no animal population acting as a reservoir of infection. (iii) There is an effective vaccine.
  ☐ Despite the discovery of an effective vaccine in 1796, it took nearly 200 years to bring about the eradication of smallpox. What were the three factors that led to successful eradication in the 1970s?
  ■ (i) The production of a heat-stable, freeze-dried vaccine in the 1950s. (ii) The development of a new vaccination technique (the bifurcated needle) in the 1960s. (iii) The World Health Organisation's provision of finance, administration and logistical support to many poor Third World countries.
  ☐ The WHO campaign abandoned a mass vaccination approach. What approach did it adopt instead?
  ■ Surveillance and containment of individual cases.

The treatment of infectious diseases

Another way of preventing the spread of infections is by rapid treatment of the disease before it can be transmitted to uninfected people. This, for example, is the main line of defence against gonorrhoea, an infection which does not produce any long-lasting immunity in its victims. One consequence of this is that people can contract a second attack of the disease within a few weeks of a previous one. For this reason it seems unlikely that it will be possible to produce an effective vaccine against gonorrhoea.

The main aim of drugs used in infectious diseases is to harm the infecting organism without harming the human host. The drugs which act against bacteria are known as antibacterials and the most important groups are the sulphonamides and the antibiotics. They work mostly by interfering either with the metabolism of the microorganism or with its structure (for instance, they may prevent the formation of the bacterial cell wall). Although there are many effective sulphonamides and antibiotics acting against bacteria, there are hardly any effective antiviral agents.

  ☐ Why do you think this is? Remember that viruses spend most of their time inside host cells.
  ■ The nucleic acid of viruses actually joins with strands of host DNA so that they become *part* of the host cell. Since viruses do not metabolise for themselves, but use host metabolic systems, it would be difficult to damage viruses without interfering with the host cells' metabolism.

Drugs that are used successfully against protozoa and helminths also tend to damage the cells of the host. This is because these organisms have a cell structure and metabolism that are similar to those of humans. In practice, this is more of an inconvenience than a major problem, as the side effects that anti-protozoan and anti-helminthic drugs produce tend to be only minor.

  ☐ Why do you think that antibacterial drugs affect host cells much less than drugs acting against protozoa and helminths?
  ■ The cell structure and metabolism of bacteria are different from those of human cells (for example, bacteria have no nucleus), so it is relatively easy to find drugs which poison bacteria but not humans.

Since the development of the first antibiotic, penicillin, in the 1940s, many different types have been produced. They have proved extremely useful in the treatment of bacterial infections and have removed much of the dread of infectious illnesses from everyday life. However, some problems associated with antibiotic treatment remain.

  ☐ One such problem was mentioned in the discussion of fungal infections, earlier in this chapter. Can you recall what it was?
  ■ Each antibiotic tends to kill several different bacteria, not only the one causing the disease being treated. This means that the normal balance between different commensal microorganisms is disrupted. Thus antibiotic treatment may lead to the overgrowth of one particular organism, such as the fungus, *Candida*, in the vagina.

Another problem that has emerged in the thirty years of widespread use of antibiotics is the problem of *bacterial resistance*. This is an example of co-evolution, discussed in Chapter 11. In any one species of bacteria, the different strains have slightly different genes and slightly different ways of metabolising. This means that some may be less affected by an antibiotic than the others. Inadequate doses of an antibiotic may kill off all the most susceptible bacteria, leaving the resistant ones to multiply and infect someone else. The infection they cause will not be very susceptible to the antibiotic. If this process of 'selective breeding' goes on long enough, the bacteria will become completely resistant to that particular antibiotic. The widespread and often indiscriminate use of antibiotics has meant that many strains of bacteria are now resistant to the common, cheaper antibiotics such as penicillin. For instance, thirty years ago, all strains of the gonococcus were sensitive to penicillin. Resistant strains first appeared in Asia in the 1960s and have now spread all over the world. By 1976 they had reached London and by 1982 2–3 per cent of cases of gonorrhoea were caused by penicillin-resistant bacteria. So far, there has been such rapid production of new antibiotics that there have always been alternatives available in cases of resistance. However, this can mean that the effective treatment of a bacterial infection is delayed if the 'wrong' antibiotic is tried first.

For both this reason and the disturbing effect antibiotics have on normal commensals, the unnecessary use of antibiotics is discouraged, particularly their routine and inappropriate use in mild infections, such as colds and sore throats, which are usually caused by viruses. However, it is also clear some people put considerable pressure on their doctor to prescribe antibiotics for viral infections, an action that may be justified for the prevention of so-called secondary infections (bacterial infections of tissues that have already been damaged by the virus).*

Summary

With improved standards of living, infectious diseases are no longer a common cause of death in industrialised countries. People may be so used to regarding this as an amazing success story that they may forget the degree to which infectious diseases still affect our lives. Much absence from work in the UK is caused by infections of the nose and throat, by colds and coughs. Although mortality from infectious diseases has declined, the incidence of minor illnesses remains high.

Most of the time humans live in harmony with microorganisms, and remain unaware of their presence on the skin, in the mouth, and in the intestine. Indeed, we become aware of them only when the delicate balance between the microorganisms' virulence and our own resistance is disturbed by some factor and we fall ill. Even when that happens, in the majority of cases the body is able to re-establish the balance and restore health, but as you have already seen in Chapter 11, this ecological balance is not static. It is continually evolving. We must therefore guard against complacency and over-confidence in assuming that we need no longer worry about the spectre of infectious diseases in industrialised countries, for an accidental or unforeseen alteration to the physical environment has the potential to so disturb the harmony with microorganisms as to threaten human survival. In Chapter 13 you saw how we need to handle the inanimate physical environment with care if we are to improve and preserve our health. Just as much care is required when considering the impact of our social, agricultural and industrial development on the animate environment, with which we share this planet.

---

* A discussion of the influence some patients have on doctors appears in the article by Helman in the Course Reader (Part 1, Section 1.2).

Objectives for Chapter 14

When you have studied this chapter, you should be able to:

14.1 Describe the six categories of organisms that cause disease, the factors that normally prevent infections occurring and the mechanisms by which infective organisms overcome those factors.

14.2 Explain how infections may be caused by commensal organisms and describe the various means by which infectious organisms may be introduced into the human body — airborne; skin contact; food and water borne; and insects.

14.3 Describe how a person's immunity to infectious disease can be raised by vaccination and the factors that limit the effectiveness of this means of disease control.

14.4 Discuss the successes and failures of treating infectious diseases with drugs.

Questions for Chapter 14

1 (*Objective 14.1*) *Streptococcus pyogenes* may be found as a commensal in the throat of many people. It may also cause severe inflammation of the throat. Can you explain how this apparent contradiction arises?

2 (*Objective 14.2*) What is the main reason why typhoid and cholera are not endemic in the United Kingdom, whereas colds and influenza are? Can you think of two other features of these diseases that account for the fact that outbreaks of typhoid and cholera are easier to control than those of colds and influenza?

3 (*Objective 14.3*) Why have worm infestations proved so difficult to control in Third World countries by any of the three standard approaches — interrupting transmission, vaccination, and treatment?

4 (*Objective 14.4*) During or after a course of antibiotics, people often get diarrhoea. Why do you think that is?

15
Living
with
ourselves

So far in these chapters on disease, we have considered humans as being little, or no, different from other organisms. We have ignored the fact that humans have feelings and react emotionally to the world around them. In this chapter we shall present some of the evidence for the ways in which human feelings and psychological responses may affect health. We hope to show why the links between psychological matters and health are by no means universally accepted and how difficult it often is to demonstrate such connections unequivocally.

Traditionally, the human passions have been associated with falling ill, as you saw in the quote from Thomas Wright in Chapter 11. In the nineteenth century it was widely thought that someone's emotional state contributed to their susceptibility to tuberculosis and to the course of this illness. Some physicians recommended love affairs for their curative properties, while others advised the avoidance of passion since this would exacerbate the disease. Nowadays, popular opinion associates other diseases with the emotions, such as heart attacks with stress and hard-driving personalities. The medical profession also links certain illnesses with emotional state. For example, the bouts of pain associated with duodenal ulcers often occur for only short periods of a few weeks or months, with quiescent times in between. These periods of pain are said to be associated with anxiety, hard work or 'stress'. Similarly, people with certain skin complaints are often said to have 'anxious personalities' and to get recurrences of their rash when they are finding life difficult. One must, of course, be careful in placing too much confidence in these traditional associations. Anxiety may be a consequence, rather than a cause, of a disease. Alternatively, any causal association may be indirect rather than direct.

☐ What factors other than anxiety do you think might cause exacerbations in the symptoms of duodenal ulcer during times of over-work?

■ A person who is anxious or busy may smoke more cigarettes, or eat irregular meals, both of which may play a part in the exacerbations of pain from a duodenal ulcer.

In other words, anxiety and pain may be *indirectly* associated via a factor such as irregular eating habits. It is very easy to suggest that a factor, loosely termed *'stress'*, contributes directly to the development of disease without taking into account possible intermediary factors, such as behavioural changes. It is also easy to hide behind such vague terms as 'stress', which has been used in different contexts to mean both physical and mental pressure, unpleasant environmental conditions, or the feeling of strain a person may feel under such pressures. In the rest of this chapter we shall try, when discussing experimental evidence, to make it clear exactly what the experimenters were measuring, and therefore exactly what they have shown to be associated with disease.

Physiological responses

Before we actually start to look at how a person's feelings and personality may predispose them to illness, we should first look at how they may affect the normal functioning of the body.

☐ You have already met at least one situation, in the discussion of the autonomic nervous system, where emotional state is reflected in the biochemistry of the body. Can you recall what that was?
■ During fear, anger, or sudden anxiety, the sympathetic nervous system is active and secretes adrenalin at nerve endings and into the bloodstream. This chemical causes decreased blood flow to the skin and gut, and increased blood flow to the muscles, increase in sweating, dilation of the pupils of the eyes, increase in the diameter of the bronchi, and an increase in heart rate — (the 'fight or flight' reaction).

Clearly, being angry or afraid alters the way your body works, but this response is short-lived. What about longer-term effects?

□ What is the principal system responsible for the long-term control of body functions?

■ Long-term control is usually via the hormonal system.

Changes in the levels of particular hormones mean changes in many different parts of the body. How far are hormone levels affected by a person's feelings? The best evidence comes from hormones produced by the outer layer or cortex of the adrenal glands. These hormones, the corticosteroids, are concerned both with the long-term maintenance of blood pressure and with the water and salt contents of the body. Corticosteroids are also known to affect the immune system: high levels depress the immune response.

Many experiments on animals have shown that corticosteroid levels change in response to environmental factors. When a loud noise or a mild electric shock is delivered to a rat, it is found that the corticosteroid levels rise immediately. Overcrowding also seems to cause high corticosteroid levels. If laboratory rats are kept in quiet, uncrowded cages, their average corticosteroid levels are about 10 per cent of those of rats kept in a noisy, crowded environment. Not surprisingly, such 'stress factors' in the environment have been shown to decrease the immune response in rats and mice, an effect that might be expected to increase their susceptibility to infectious diseases. It is not certain whether all these effects are mediated by corticosteroids since other hormone levels may be affected by noise and electric shocks. These too are thought to affect the immune response.

One study tried to look more deeply into the effect of an electric shock on rats. Two groups of rats were given a mild electric shock. One group was able to stop the shock and escape by turning a wheel in their cage, the other group suffered identical shocks, but were unable to stop them or escape. Twenty-four hours later, immune responsiveness was tested in each group and was found to be lower in the group that had suffered inescapable electric shocks. The group that could escape had immune responses similar to control rats that had not undergone any electric shocks.

□ What do you think that this suggests?

■ The effect of a stimulus is modified by the action that can be taken in response to it.

Corticosteroid levels have also been measured in humans in stressful conditions. One such study among students compared their corticosteroid levels during academic examinations and found them to be considerably elevated above normal levels. Another study in the USA investigated parents of children with leukaemia. The parents were interviewed and their reactions to their children's illness assessed. They were grouped into those who were much

distressed and admitted to feelings of despair and those who seemed less distressed and despairing. Their corticosteroid levels were then measured. The least distressed group had normal levels. The very distressed group had abnormally high levels.

Accurate methods of assessing hormone levels and testing the immune response have only recently been devised. It would seem, from the experiments that have been done so far, that there is a link between emotional responses and hormone levels and, hence, to the day-to-day functioning of the human body. If this is so, it may be that there is a connection with disease. There is, however, no conclusive evidence on this point.

Pathological responses

The connection between feelings and disease has long been recognised by writers and poets, doctors and, most significantly, the general public. It is only in recent times, however, that there have been attempts to discover a scientific basis for such widely held beliefs. There have been two main types of research: that which tries to examine the external environment and relate it to illness, and that which looks at the way different people react to certain situations and examines whether different types of response are associated with differing susceptibility to various diseases.

One example of the former type of study looked at the connection between asthma and the family. You may remember that asthma may occur in people who are allergic to some environmental antigen, but not all people who get asthma can be shown to be hypersensitive to environmental antigens. About 30 per cent of people with asthma seem to get attacks when they are upset or anxious. There has been a suggestion, among some doctors and psychologists, that children with this sort of asthma have parents who are 'overprotective' and anxious. There are problems with this sort of impression: who is to say that parents do not become 'overprotective' because their child has asthma, rather than contributing to the asthma by their overprotectiveness? The study aimed to see whether the frequency of attacks in children was related to their parents' presence. The children who took part had no known allergies, but there was subjective evidence that emotional upset precipitated attacks of asthma. These children were studied under three conditions: at home with parents, then at home with baby-sitters while the parents were away and again at home with parents. The results are shown in Figure 15.1.

□ Did the children have fewer or more asthma attacks during the baby-sitter periods of separation?

■ Fewer. They also required less medication.

A parallel study, of asthmatic children in whom there was a known allergen that precipitated attacks, found that their attack rate was unaffected by separation. It would seem, therefore, that parents had something to do with the attacks in the non-allergic group and that some symptoms could have been, though not necessarily, due to the child's emotional response to its parents.

That study looked at the effect of the family. Other researchers have tried to look at the effect of stressful stimuli in the environment. To do this they have analysed events that occur in people's lives, such as getting married, the death of a close relative or the loss of a job, and ranked these 'life events' according to how much change in lifestyle they are estimated to cause. Each event is given a score: the higher the score, the more significant the change in lifestyle the event supposedly causes. Subjects are asked to list the events that have occurred in a set period of time, their scores are added up, and the total gives a measure of the change in their lives during the set period. After this they are asked to state what illnesses they have had in the periods studied. Several such investigations have shown an association between events requiring social adjustment and the incidence of illness. One study claimed that major life changes mean a greater susceptibility to illness for as long as two years after the time of change.

    ☐ Can you think of any problem in the method of collection of these data?

    ■ It is possible, or even likely, that people tend to notice, or to remember, illnesses more if they occurred during, or soon after, times when they were busy with changes in their lives.

This difficulty can be avoided by doing prospective studies. One such experiment asked 84 young American doctors for details of life events in the previous eighteen months. Nine months later they were asked about their health in the intervening time. Of those with high 'life event' scores, 49 per cent reported illness compared with 25 per cent of those with medium scores and 9 per cent of those with low scores.

    ☐ What objections might be made about the validity of these results?

    ■ It is possible that different people fill in the questionnaires with differing amounts of care. Conscientious people may take more time to remember their life events and *also* what illnesses they have had, and hence remember more of both.

Nevertheless, there is accumulating evidence of associations between life events and the incidence of illness. On the basis of current research it seems likely that stressful events can precipitate overt illness. This association is seen despite the fact that different people respond differently to similar upsetting events: some take them in their stride, whereas others are seriously disturbed by them. This brings us to the second type of experiment, where people's *response* to

situations is assessed. Studies of this sort have tended to concentrate on two diseases, coronary heart disease and cancers. Here we shall concentrate on the studies that have tried to correlate people's personality and their susceptibility to cancers. There are two main questions to ask: do cancers occur more in people who react to life in one particular way, and does a person's response to having cancer alter the course of their disease? Neither question is easy to answer, and results have often been conflicting.

Two studies, one in Britain and one in Australia, have followed men suffering from depression and have found more cancer deaths than would have been expected. However, two other studies, in Britain and Finland, showed no such increase in incidence. Even if such an increased incidence were to be found, it would be necessary to examine what drugs the subjects had taken for their depression since these, rather than the depression itself, might be a factor in susceptibility to cancer.

In general, these studies have been retrospective, that is, people with cancer have been interviewed and their personality assessed. This is usually done by asking them how they would respond to certain situations. The cancer group is then compared with a control group of people of the same sex and age, but who have a different illness. Such studies as these have suggested that: men with lung cancer had fewer opportunities to express emotion compared with men with other lung disease; a tendency to suppress anger

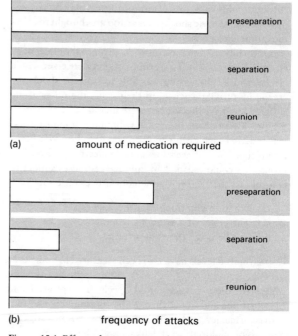

(a)   amount of medication required

(b)   frequency of attacks

Figure 15.1 Effect of separating asthmatic children from their parents on (a) their need for medication and (b) their frequency of attacks. (From Purcell, K. *et al.* (1969) p. 153.)

was associated with the occurrence of breast cancer in women less than 50-years old; women who were awaiting the biopsy (removal for microscopic study) of a breast lump who showed low anxiety were more likely to have cancer than those who were highly anxious; repression of feelings occurs more frequently in men with cancer than among controls. However, it must be pointed out that other studies have failed to show an association between personality and cancers.

□ What is the problem with *retrospective* studies such as these?

■ You can never be sure that the cancer resulted from the personality characteristic. It is possible that the characteristic is a result of having cancer: for instance, people with cancer may learn to suppress their anxieties.

One prospective study of American students who entered medical school between 1948 and 1964 is still going on. So far it has shown a slight association between lack of closeness with parents and the development of cancers in later life. Other prospective studies supporting this finding have involved only small numbers of people and are not statistically significant. The concept of a cancer-prone personality is not yet generally accepted.

Several 'alternative health' groups maintain that it is possible to alter the course of cancers by the 'correct' attitude of mind. The evidence for this statement is conflicting, and there are often methodological inadequacies in the studies quoted. For instance, some compare survival times after the diagnosis of cancer is made, without estimating the extent of the cancer at the time of diagnosis. Three investigations are worth quoting here. One American study followed up people with several types of cancer and devised a method of comparing expected survival time with actual survival time. The authors found that long survivors had closer personal relationships, were less distressed, found their doctors more helpful and complained less than those who survived for only a short time. In direct contrast to this, a study of women who had breast cancer that had already spread to other parts of the body showed that those who lived longer showed more emotional distress, and more negative attitudes to their doctors. Finally, in a prospective study at King's College Hospital, London, women with early breast cancer were assessed psychologically by interview three months after mastectomy (removal of the breast), and were then assessed medically at regular intervals for five years. The women were divided, according to their response to having cancer, into four main groups: denial, fighting spirit, stoic acceptance, and helplessness/hopelessness. Their psychological response three months after the operation was found to be related to their condition five years later: people in the fighting spirit or denial groups did better than those in the stoic acceptance or the hopelessness groups.

Not surprisingly, with such conflicting results, the case remains unproved. As we have tried to indicate, this whole area of research is fraught with difficulties. Nevertheless, as the links between the nervous system, the hormonal system, the immune system and people's emotional responses are worked out, the situation should become clearer. Work in this field is proceeding apace, and there may well be more definite results in a few years. Whether this will be useful or not is a different matter. After all, what can be done if it is discovered that people lacking close parental ties are more susceptible to cancer? On the other hand, if it is shown that a certain type of response to stressful events tends to be associated with disease, then it may be possible to help people to alter that response. Such an approach has been used in the treatment of hypertension (high blood pressure). In some people, blood pressure rises in later life and this predisposes them to strokes. For this reason, they are often given drugs to lower their blood pressure.

Hypertension is thought to be stress-related in humans and can be caused in laboratory animals by overcrowding and noise. This has encouraged people to investigate whether blood pressure can be lowered by altering people's response to events in their lives. One British doctor, Chandra Patel, has produced a treatment package for hypertension, using control of physiological factors by relaxation training to minimise a patient's adverse response to stress. Patients suffering from hypertension are first trained to control physiological factors associated with stress using relaxation and meditation techniques. In the next stage patients are taught to recognise in their environment the factors that they find stressful, and then to relax when faced with these factors. In this last stage, the patients use a system of red dots placed at points where stress may occur — on a watch, for instance, for those for whom looking at their watch is associated with hurrying, or on car keys for those who experience tension in driving. When the patient sees a red dot, she or he makes a conscious effort to relax. Eventually the relaxation becomes part of the patient's routine. This treatment package takes only six weeks and has produced significant falls in the average blood pressure of recipients compared with a control group who attended the clinic but did not receive stress management training. The effects were maintained when patients were followed up three months later. This can be a useful way of treating people with hypertension. It does, however, require considerable motivation from the patient and, as well, a willingness to change the way they respond to situations — not everyone wants to do this, even if their method of coping is said to be 'unhealthy'. However, if such methods do prove to be useful in a wide variety of illnesses, some people may prefer them to taking drugs or undergoing other conventional medical treatments.

Summary

In Chapters 13 and 14, you saw how important it is, for continued health, to maintain a state of balance, or homeostasis, with material factors in the external world, be they animate or inanimate. Equally important to health is the ability to maintain internal homeostasis in the face of psychological factors, such as stress. Studies of short-term stress, physiology and life events all suggest that there is at least a plausible link between psychological factors, physiological responses and persistent changes that might result in pathological states. In many ways, knowledge and understanding of this area is primitive compared with that of the physical factors discussed in previous chapters. (The references for the research studies mentioned in this chapter are given in full at the end of this book.) However, it seems clear that, in industrialised countries, the influence of the external world on people's psychological state may be at least as, if not more, important to their state of health than chemicals and radiation, viruses and helminths.

Objectives for Chapter 15

When you have studied this chapter, you should be able to:

15.1 Describe some of the physiological and pathological responses of the human body to psychological factors such as emotions and stress.

15.2 Understand the difficulties in demonstrating the connections between a person's psychological state and the occurrence of ill-health, and be able to give some examples of such studies.

Questions for Chapter 15

1 (*Objective 15.1*) A champion motor-cyclist has described how the frequent asthmatic attacks he suffered throughout childhood vanished when he left home and took a job that involved spending much of his time travelling around Europe. He also reported that, when he had suffered asthmatic attacks at home, the only effective 'treatment' was being comforted by his mother. What features of this story support a psychological aspect to his ill-health?

2 (*Objective 15.2*) A study in Boston in 1962 found that routine throat swabs from children showed streptococcal infections more commonly at times of family crisis than at other times. Does this show that an increase in stress causes an increase in streptococcal throat infections?

16
A multiplicity of causes

During this chapter you will be referred to the articles in the Course Reader on 'Ethical dilemmas in evaluation' (Part 3, Section 3.6).

So far, our discussion of diseases has considered those conditions in which the proximate cause is well-recognised. This chapter is concerned, as its title implies, with those diseases in which the proximate cause has not yet been identified. As you will see, it seems likely that for some of these diseases there may not be a single proximate cause. Instead, it may be necessary for several causative factors to coincide or accumulate within an individual for the disease to occur. Such a *multifactorial model* of cause is currently attracting much interest and the epidemiological methods that are frequently employed in research into causation are well-suited to the investigation of such a theory.

The diseases covered in the chapter account for the majority of deaths, illnesses and disabilities in industrialised countries today. This is not, of course, a mere coincidence. If the causes of these conditions were understood better, it is unlikely they would continue to make up such a high proportion of the burden of disease. As with previous chapters, we can consider only briefly a few of these conditions. We have, however, devoted considerably more attention to the subject of cancers, in view of the significant changes that have occurred in recent years in theories and evidence as to the causal mechanisms involved.

Neural-tube defects

Neural-tube defects form a class of disorders that arise from congenital faults in the development of the brain and the spinal cord and their surrounding tissues. In the severest form, the brain does not form at all, a condition known as anencephaly. Anencephalic babies are usually stillborn, or die shortly after birth.

In the less severe form, spina bifida, the abnormality is mainly at the lower end of the spinal cord. Normally, the spinal cord forms at an early stage in the embryo's development and then the bones of the spine condense around it. Figure 16.1 shows both the normal result and spina bifida. In spina bifida the bones do not unite over the

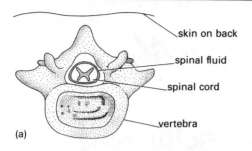

(a)

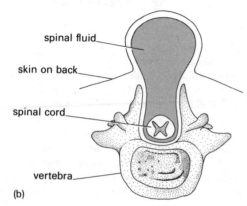

(b)

Figure 16.1 Cross-section of (a) a normal spine, and (b) a spine affected by spina bifida.

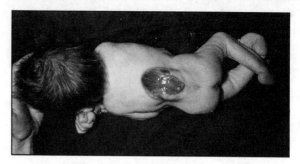

Figure 16.2 Infant with spina bifida.

spinal cord, so there is a gap. Sometimes the gap is small and the skin forms normally over it. The defect may, however, be more severe so that even the skin does not form over it and the spinal cord is actually exposed to the air (Figure 16.2). Often in these extreme cases the spinal cord itself is malformed and the nerves to the lower limbs, large bowel and bladder, are severely affected.

Although these severe defects can be repaired surgically to some extent, the affected children suffer further developmental problems. This is because the posture, or shape, of the body is partly produced by the pull of muscles. Since some of the nerves to the back and leg muscles are abnormal, some muscles are weaker than others and their relative pull on the skeleton is abnormal. As a result, deformities of posture may arise, particularly curvature of the spine as the child grows.

Another problem which frequently affects children with severe spina bifida is a disorder of the drainage of fluid that bathes the spinal cord and brain (the cerebrospinal fluid). This fluid is formed from specialised capillaries in the cavities (known as ventricles) in the centre of the brain. It passes to the outer surface of the brain and circulates around both the brain and spinal cord, before being

reabsorbed back into the veins within the skull. If the draining pathway is blocked, fluid accumulates under increasing pressure. In young babies the bones of the skull are not yet fused, so the head tends to enlarge under the increased internal pressure, a condition known as hydrocephalus, or 'water on the brain'. Left untreated, this condition will lead to progressive damage to the brain.

It is possible by surgical operation to close the skin and bone defect in severe spina bifida and, in hydrocephalus, to insert drainage shunts within the skull to allow the free passage of cerebrospinal fluid. Some doctors, however, think that, because of the problems that continue to occur throughout life, even after such surgery, severe spina bifida should not be treated.

☐ What do you think will happen if the skin defect is not repaired and the spinal cord is left open to the air?
■ Bacteria will enter and breed in the cerebrospinal fluid, causing inflammation of the tissues around the spinal cord (meningitis). If untreated, this will prove fatal.

Non-treatment means that about 80 per cent will die before the age of 13: treatment means a life of predictable complications and disabilities.

Although the cause of spina bifida is unknown, its occurrence is known to be associated with several factors. Couples who have had one child with spina bifida are more likely to have a second (1 in 20) and the risk rises to 1 in 10 after two affected children. There is an unexplained geographical variation in incidence with higher rates in Northern Ireland than in the rest of the UK. There is also considerable support for the idea that spina bifida may result from inadequate vitamins in the maternal diet.*

---

* Evidence for this can be found in a collection of articles in the Course Reader on 'Ethical dilemmas in evaluation' (Part 3, Section 3.6).

Cardiovascular disease

In young children the lining of the arteries is smooth and white, providing low resistance to the passage of blood. Figure 16.3 shows the inside of the aorta (the main artery leaving the heart) of an elderly man. The surface could hardly be described as smooth and white. Its bumpy, irregular appearance is typical of the changes in arteries seen, to some extent, in all adults in the UK. This condition is known as atherosclerosis because the artery walls contain hard plaques filled with fatty deposits that resemble porridge (athero = porridge, sclerosis = hardening). It is these fatty deposits in the arterial wall that are responsible for coronary heart disease and strokes, which are a major cause of death in industrialised countries.

How does this distortion of the arterial wall happen? The first changes seem to occur in late childhood when long, thin, fatty streaks appear on the inside of the arteries.

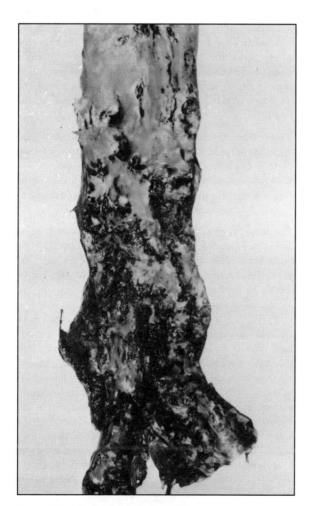

Figure 16.3 Atherosclerosis inside the aorta.

The streaks have a smooth surface and do not seem to cause any harm. They may not be permanent, but their numbers do seem to increase with age and they may be the first step towards the more serious damage shown in Figure 16.3. The plaques consist of fats, in particular cholesterol, and fibrin (a protein normally produced when blood clots form). There are two main theories as to how these plaques arise. The first suggests that fats pass from the blood into the arterial wall where they provoke a proliferation of scar tissue. The second suggests that blood clots occasionally form on the arterial wall because of turbulence of blood flow and that these clots get covered by the cells that line the artery and are thus incorporated into the arterial wall. There they degenerate into the fat and fibrin found in the plaques. What is certain is that once these plaques have formed their roughened surfaces form a focus for the formation of thrombi (blood clots in the blood vessel: the singular is thrombus).

☐ What effects do you think atherosclerosis might have on the functioning of the cardiovascular system?
■ The main effect is that the plaques narrow the arteries so that insufficient blood passes through to the tissues.

In addition, if a piece of plaque or thrombus breaks off (an embolus), it may travel down the increasingly narrow artery and completely block it. Thrombus formation on a plaque can also completely block off an artery. This restriction in blood flow is known as ischaemia.

Atherosclerosis occurs particularly in the major artery leaving the left ventricle of the heart, known as the aorta, and in its first two small branches, the coronary arteries which supply oxygenated blood to the heart (Figure 16.4). Disease resulting from an inadequate supply of oxygenated blood to the heart muscle by the coronary arteries is known as coronary or ischaemic heart disease.

The coronary arteries are quite small and are therefore easily blocked. Partial blockage usually allows enough blood through to meet the needs of the muscle when the heart is beating slowly. During exercise, however, the heart beats faster and therefore needs more oxygen. If the blood flow is inadequate during exercise, the heart muscle will receive insufficient oxygen. This will cause cramp in the heart muscle, which is felt as pain in the chest or arm. The pain, known as angina, gradually goes away once the person stops exercising and the heart rate returns to normal. In this context, exercise can mean as little as running for a bus or climbing stairs.

☐ What will happen if a coronary artery becomes completely blocked off?
■ The heart muscle that it supplies will be completely deprived of blood and the muscle cells will die.

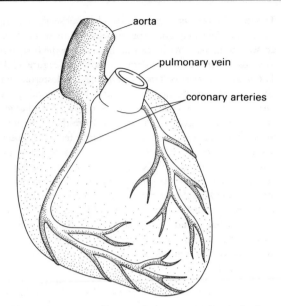

Figure 16.4 The two coronary arteries supplying the heart muscles.

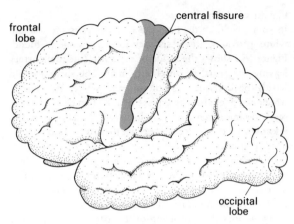

Figure 16.5 Side view of cerebral cortex showing an area that is commonly affected in strokes.

This is known as a heart attack or myocardial infarction. It may also be referred to as coronary thrombosis, though strictly speaking this means that the blockage is due to a thrombus (clot) rather than some other mechanism.

The effect of a dead patch of muscle in the heart depends on its extent and location. If the area of the infarction is large, then the heart is no longer able to pump blood round the body adequately. If the infarction is on the left side of the heart, the back-pressure of blood in the lungs leads to fluid passing out of the blood vessels into the surrounding lung tissue. This results in breathlessness. If the infarction occurs in the right side of the heart, pressure will build up in the circulation in the rest of the body, so that fluid accumulates throughout the body, particularly around the ankles. This condition is known as heart failure.

If an infarction affects only a small patch of heart muscle, then the person will experience an episode of chest pain and feeling ill, and will probably recover. They will, however, be left with a small patch of scar tissue in their heart. If the infarction includes part of the heart tissue that initiates the contraction of the heart or conducts the message around the heart muscle, then the heart may develop a disordered rhythm (a cardiac arrhythmia) and may even stop. This is one common cause of death in heart attacks.

The other organ that is commonly affected by atherosclerosis is the brain. The effects depend on which areas of the brain suffer from a shortage of oxygen and nutrients. Small areas of the brain may actually die. This is one cause of confusion, loss of memory and intellectual impairment in elderly people, a condition known as senile dementia.

Sudden, complete blockage of arteries in the brain, or haemorrhage from blood vessels in the brain weakened by atherosclerosis, may lead to the sudden death of large areas of brain tissue. Such an occurrence is known as a stroke or cerebro–vascular accident. The effect will depend on the area involved.

☐ What effect might you expect from damage to cells on the left side of the brain in the areas shown in Figure 16.5? (You may need to refer back to Figure 6.8.)

■ This area of the brain is responsible for initiating movements in the right side of the body, so damage to it will cause weakness on that side of the body.

The part of the brain that controls speech is near this area and strokes affecting movements on the right side of the body often cause difficulties of speech. For instance, the person may know what he or she wants to say, but be unable to find the right words to say it, or may use the wrong word without realising.

Finally, atherosclerosis may affect arteries in the rest of the body and cause ischaemia, particularly in the feet and legs. This is known as peripheral vascular disease. It is especially common in diabetics, who often have severe atherosclerosis. If the blood supply to the skin is principally affected, infarction of the skin may occur leading to chronic ulceration. Sometimes toes, or even the whole foot or leg, have to be amputated because of a poor blood supply. If the arteries to the muscles of the leg are affected, their blood supply will be inadequate for any exertion and the muscles may develop cramp whenever they are used. Some people with this condition can walk only a few yards before developing severe pain in their legs.

Though the proximate cause of both atherosclerosis and

the resultant diseases is not known, there is a considerable knowledge about their associations with several *distal* factors. Ischaemic heart disease (IHD) is associated with: increasing age; a family history of premature IHD; being male (at younger ages); dietary factors such as fat consumption; raised blood pressure; diabetes; obesity; physical inactivity; smoking; psychosocial factors such as personality and response to stress; winter season and cold weather; softness of the water supply; being a heavy drinker and, paradoxically, being a teetotaller. In the light of so many distal associations, it is generally believed that there is no single proximate cause, but that IHD has a multifactorial cause.

Hypertension

Another condition that affects the circulation of the blood is hypertension, or high blood pressure. You may remember from Chapter 7, that the pressure in an individual's arteries varies with each heartbeat, from a pressure of about 120 to 80 mm of mercury (in a healthy young adult). In addition, blood pressure is altered in the short term by physical exercise and emotional experiences. At any age there is considerable variation in blood pressure between different individuals, but in all industrial societies the average blood pressure tends to increase with age, up to about the age of 75. In a person over 70-years old, the average diastolic blood pressure is 95 mm of mercury.

There are many gradual changes in the body that occur with age, so is the rise in blood pressure of any importance? Epidemiological studies suggest that it is. Although a moderate increase in an individual's diastolic pressure rarely causes any symptoms, it is associated with an increased risk of mortality, particularly from strokes, coronary heart disease and kidney disease (Figure 16.6).

As with coronary heart disease, the proximate cause of

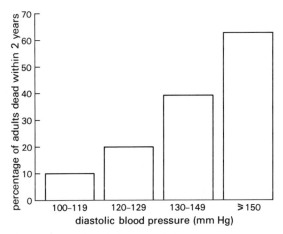

Figure 16.6 Relationship between diastolic blood pressure and deaths within two years in middle-aged adults.

hypertension is unknown. Several factors, however, are known to be associated with an increased risk of developing hypertension, including obesity, a family history of hypertension, soft drinking water, stress, heavy drinking and high salt consumption. The idea that salt consumption may be important dates back to the ancient Chinese, who suggested, 4 000 years ago, that people who engorge themselves on salt develop apoplexy (stroke). It is thought that an almost salt-free diet leads to a fall in blood pressure. Of these factors, only the increase in salt intake can really explain the rise in blood pressure with age in European populations. The suggestion is that young people can compensate for the excessive salt intake but, as they get older, this compensation becomes less efficient and their blood pressure slowly rises.

Degenerative brain disease

The nerve cells in the brain do not divide at all after birth. Nerve cells die throughout life and therefore there is a gradual decrease in their number. The average weight of the brain of a 90-year-old is 1 360 grams compared with 1 500 grams for a 30-year-old. As well as this loss of cells, there is an accumulation of abnormal proteins both inside and outside the cells. In addition, there seem to be defects in the synthesis of transmitter substances in the brain. These degenerative changes in the brain may result in a condition known as dementia, in which there is a progressive and irreversible deterioration of brain function.

These changes occur to some extent in everyone, but in some people they start early and become severe. Degeneration of brain cells and the effects of atherosclerosis on the brain are the two major causes of dementia. About 700 000 elderly people in the UK are affected to the extent that they cannot look after themselves. With the increasing numbers of very old people in the population, dementia is an increasing problem. One particular type of degenerative brain disease is known as Alzheimer's disease, the cause of which is unknown. There have been suggestions that it is due to an infective agent, such as a slow virus: that is, a virus that invades the body and lies dormant for many years before causing any disease. Others have suggested that it results purely from a speeding up of the normal ageing process in cells. Alzheimer's disease is currently attracting considerable research. At present, all discussion as to its cause contains more speculation than fact.

Degeneration of joints

Many people, as they grow older, complain of stiffness and pain in their joints, particularly in their knees, hips and hands. You will remember from Chapter 3 that the ends of bones, where they articulate with other bones, are covered with a smooth, tough layer of cartilage that is capable of withstanding considerable force. As people get older, the

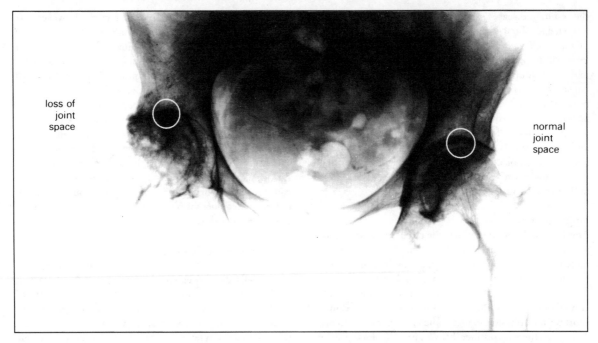

loss of
joint
space

normal
joint
space

Figure 16.7 X-ray of pelvis showing loss of the joint space in the right hip due to osteoarthrosis.

cartilage-producing cells divide less often and the substances they produce become less able to withstand the forces to which they are subjected. The cartilage starts to break down and surface layers may flake off. This change is found in nearly everyone as their age increases. In some people, the end of a bone actually becomes uncovered and grates against the other bone in the joint. The bone ends become dense and thickened in response to this stress and their surfaces, not strong enough to withstand the friction, become grooved and furrowed. At this stage people often experience pain and stiffness in the affected joints. The condition is known as osteoarthrosis. This used to be called osteo-arthritis, meaning *inflammation* of the joint, but the changes seen are not the same as those that occur in acute or chronic inflammation, so the name has been changed. Figure 16.7 shows an X-ray of the pelvis of someone suffering from pain in the right hip.

☐   What differences can you see between the two hip joints? (This X-ray has been printed as a positive image, so bone appears grey and dense areas of bone look even darker.)

■   The right hip joint shows the changes which are typical in osteoarthrosis: the space between the two bones is narrow, because the cartilage has disappeared, and the actual surfaces of the bones are dense and irregular.

In about three-quarters of the cases of osteoarthrosis that

come to medical attention, some underlying cause for the condition can be found. For instance, deformities in the bones, or unusually stressful physical activities, can put abnormal pressure on the cartilage that covers the bone ends. Occasionally the disorder is associated with a gene defect that results in an abnormality of the cartilage. Often, however, no such contributory factor is known and the disorder is presumed to be due to a severe form of the normal changes that occur in everyone's cartilage as they get older.

A different type of joint degeneration, known as rheumatoid arthritis, affects people of all ages, but is most common in older people. It is an example of an auto-immune disease and is a disease of varying severity in which the major damage is to the joints. Rheumatoid arthritis affects at least 1 per cent of people in the UK, though the incidence may be higher since it may go unrecognised in people who have only a mild form.

Rheumatoid arthritis may first become evident at any age. It tends to flare up and settle down throughout life. During the periods of acute illness, affected joints become hot and swollen and the person has a fever and feels ill. If the joints are examined at this time, it is found that they show all the changes of an acute inflammatory reaction, with dilated blood vessels, infiltration by macrophages and swelling of the membrane lining the joint (the synovial membrane), which secretes lubricating fluid. When this stage of fever, pain and swollen joints settles, the state of the joints becomes typical of chronic inflammation. The tissues are full of lymphocytes, there is a patchy overgrowth of the synovial membrane and there may be gradual destruction of the cartilage over the bone ends, with consequent deformity of the joints (Figure 16.8).

Why does this condition occur? There is probably some genetic factor involved. A study in Denmark showed that identical twins, who share the same genes, suffer from rheumatoid arthritis more often than non-identical twins, whose genes differ, and members of the general population have a higher risk of developing rheumatoid arthritis if they have a close relative with the disease. This does not exclude common environmental factors. Rheumatoid arthritis often first appears after an infection and some people think that the disease is caused by infective microorganisms that stimulate a chronic inflammatory response in the joints.

The auto-immune diseases, such as rheumatoid arthritis, demonstrate the difficulties of trying to find the proximate cause for some diseases, especially when there seems to be more than one contributory factor. Perhaps in some people there is a genetic predisposition to produce anti-self antibodies or lymphocytes and this tendency may increase with age. The actual production of a harmful immune response may need an external trigger, such as an infective organism, a chemical or radiation. Thus, many factors may contribute to the formation of the disease, with two or more needed for it actually to occur.

Cancers and benign tumours

We turn now to a cluster of diseases which give rise to a bewildering array of symptoms. Although little is known about the proximate causes of cancers, a considerable amount has been discovered about distal causes and the pathological changes that occur. For example, despite the vast amount of research that has been carried out into the cause of lung cancer, the chemical in tobacco smoke that is believed to be the proximate cause has still not been

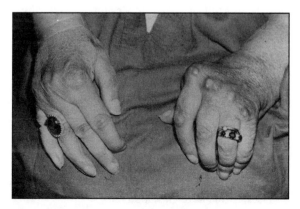

Figure 16.8 Rheumatoid arthritis affecting the hands.

identified. Much of the discussion will therefore concentrate on distal causes of cancers and, even more so, on the pathogenesis of cancers.

When discussing tumours it is necessary to distinguish between those that are benign and those that are malignant. *Benign tumours* grow as a well-defined mass of cells that are still recognisably similar to their tissue of origin. The mass is covered by a thin layer of normal cells (the capsule) from which tumour cells do not usually escape. *Malignant tumours* show a much more chaotic and disorganised pattern of growth and malignant cells exhibit varying degrees of reversion to an unspecialised state. In other words, the process of cell differentiation seems to reverse and the malignant cells look less and less like the cells in which the tumour originated. There is no capsule around a malignant tumour and the cells grow into surrounding tissues, a characteristic known as invasiveness. Some of these cells breach the walls of nearby capillaries and are swept away in the lymph or bloodstream and, hence, spread to other parts of the body. Malignant tumours are commonly known as cancers, although in medical texts they are often referred to as neoplasms (new growths).

Cancers are not equally common in all the various tissues and organs of the body. Their incidence quite closely reflects the underlying capacity of various cell types to replace 'worn out' cells by division of existing ones. You should recall from Chapter 3 that the body can be viewed as an assemblage of specialised cells in a constant and dynamic cycle of death and renewal, but that some cell types have a faster 'turnover' than others.

☐ In which tissues or organs would you expect cancers to be most common and in which the least? (You may have to deduce part of the answer from the *principles* outlined in Chapter 3, rather than recalling a ready-made solution.)

■ Cancers are most common in tissues with a relatively high 'background' rate of cell division. For example, the skin and other surfaces that are constantly worn away or sloughed off (the linings of the gut and reproductive tract, the lungs and the bladder). Blood cells are replaced throughout life and growths arising in the various types of white blood cell are relatively common. Glands, such as the pancreas, are continuously active in producing hormones or enzymes and must be able to replace 'worn out' cells, as must the liver and kidneys which filter and clean out the blood. Far less common are cancers arising initially in tissues with a low rate of cell division in their normal state: for example, muscle or bone. In adults, neurons cannot divide at all and never give rise to cancers (so-called 'brain tumours' result from proliferation of connective cells in the brain and not from neurons).

Background cell division, however, is only part of the explanation for the relative distribution of cancers in the body. You may have noticed that many of the tissues that have relatively high incidences of cancers are either exposed directly to chemicals, microorganisms or radiation in the environment (for example, in the air, food and drink) or are involved in processing or transporting the products of digestion and respiration.

A cancer begins as a mass of cells known as the primary tumour. Secondary tumours that arise by malignant cells breaking away and becoming lodged in favourable sites for growth, are known as *metastases*. A distinction is commonly made between local disease (that is, malignant, but still confined to the original location) and disseminated cancer, which has produced detectable metastases. Certain types of cancer seem to 'show a preference' for spreading to particular sites. For example, cancer of the bronchi, the main air tubes in the lung, frequently spreads to the brain and the adrenal glands, whereas breast cancers are more likely to spread to the bones and liver. No convincing biological explanation yet exists for these characteristic 'preferences'.

Cancers can also arise as a huge proliferation of detached malignant cells that are derived from white blood cells and circulate in the bloodstream (resulting in one of several types of leukaemia) or in the lymphatic system (a condition known as a lymphoma).

Benign and malignant tumours generally have very different consequences for the patient. As their name suggests, benign tumours frequently cause no trouble at all. For example, moles are areas of excess growth of the pigment cells in the skin, but they rarely cause any problem, though in rare cases, they may become malignant and spread. Similarly, fibroids, which are benign growths of muscle tissue in the wall of the uterus, may not cause any symptoms. In some cases they can cause excessive menstrual bleeding, requiring a hysterectomy. Benign tumours, however, can grow quite large and occasionally have severe consequences, especially if they are growing in a confined space, such as the skull.

□   Why do you think this could be?

■   As the tumour expands, it exerts pressure on nearby tissues, distorting them or blocking their blood supply or nerves controlling their function. The output of a gland may be restricted or parts of the tissue may die through lack of oxygen and nutrients.

Contrary to their name, malignant tumours are not *always* serious — rodent ulcers, a type of cancer arising in non-pigmented cells in the skin, are very rarely fatal. The majority of malignant growths, however, do result in death unless they are treated successfully, sometimes because of pressure effects as described above, but more usually by

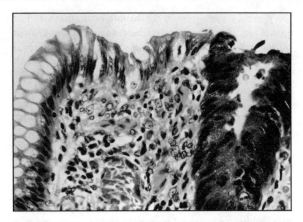

Figure 16.9 Increased rate of mitosis in malignant cells. Several groups of darkly staining chromosomes can be seen in dividing cells in the malignant epithelium on the right, whereas no such mitotic activity can be seen in the normal epithelium on the left.

spreading through the body and thereby affecting many tissues and organs.

Metastases can block blood vessels and disrupt the supply to vital organs. They can also breach vessel walls and cause bleeding, and lodge in bones, weakening their structure so that only a small stress results in a fracture. Some cancers secrete hormone-like substances that can disrupt the normal metabolism of the whole body; others secrete molecules that induce fever and sweating.

Although there are numerous differences between the cells of one type of cancer and another (for example, between lung cancer and breast cancer), five major characteristics are common to most malignant cells:

1 Cancer cells seem to lack the usual control of cell division and *divide repeatedly*. Cancers contain higher than normal frequencies of cells with visible chromosomes in the characteristic patterns of mitosis, as Figure 16.9 illustrates.

2 Malignant cells are *less differentiated* than the cells normally found in the tissue of origin: that is, they lose the form and structure characteristic of mature cells specialised to fulfil a particular function (see Figure 16.10).

3 Highly malignant cells (that is, cells with a high tendency to spread) have the ability to *migrate*, a property that most normal cells lack.

4 Malignant cells exhibit a reduced *adhesiveness* to one another and hence can detach themselves from their neighbours and 'wander off'.

5 Malignant cells display the loss of a behavioural characteristic known as *contact inhibition*, in which cells that make contact with one another stop moving. Malignant cells continue moving when in contact with other cells, an essential characteristic if the malignant cell is to escape from its original location.

Note that the properties described in points 3, 4 and 5

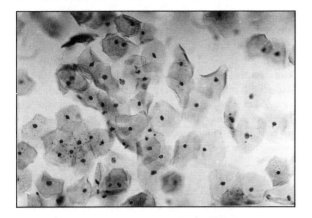

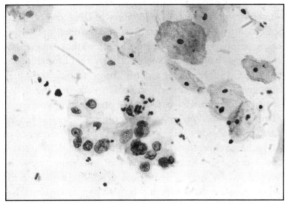

Figure 16.10 Cells from cervical smears (magnified about 200 times): (a) healthy appearance (note the regular size and shape of the cells and the relatively small nuclei characteristic of epithelium cells lining the cervix); (b) malignant features (note the very large nuclei and irregular size and shape of the cells on the left-hand side).

are essential for the invasive characteristic of malignant growths.

□ Are these five characteristics genuinely 'abnormal'? Where have you met them before in this book?

■ These characteristics are 'normal' during the development of the embryo, when undifferentiated cells multiply repeatedly and migrate by complex routes to precise locations, uninhibited by mutual adhesiveness or contact with adjacent cells.

In recent years a revolution has occurred in cancer biology as it has become accepted that malignant cells are not, as they were once thought to be, totally abnormal and 'foreign' entities. The modern view depicts them as cells which have reverted to an earlier stage of development, resembling embryonic, rather than adult, cells. *Thus they are (in most respects) biologically normal cells, but occur in the wrong place and at the wrong time.*

The biological reactions that take place in cancer cells closely resemble those of their normal counterparts: a recent study showed that 97 per cent of the protein molecules found in the cancers of experimental mice were also found in normal mouse cells. However, there *are* differences. Many cancer cells can survive low oxygen levels, by relying on anaerobic respiration far better than normal cells. They may synthesise unusual proteins: for example, enzymes that digest the substances forming a 'cement' between adjacent cells, thereby facilitating migration through nearby tissues. Cancer cells may also stimulate the growth of new blood capillaries around the tumour mass and, ultimately, within it. This is essential to ensure an adequate supply of oxygen and nutrients to the growing mass of cells. Unless an adequate blood supply is established, tumours are restricted to about 1–2 mm in diameter and consist of an inner 'core' of dying cells and an outer 'rind' of multiplying cells that receive oxygen and nutrients by simple diffusion from nearby tissues.

Cancer cells, therefore, show several significant adaptations to their new 'way of life', but these consist predominantly in *regaining* those properties lost by cells as the embryo matures. Malignancy is therefore partly an expression of the underlying properties of the *normal* cell. This realisation led to an investigation of those parts of the genetic material that are involved in issuing instructions for the cell to divide. The existence of 'silent' genes (genes in non-malignant cells that, if activated, result in transformation of the cell to a malignant state) was proposed as long ago as 1969, but confirmation has been a long time coming. They have been dubbed *oncogenes* (from the Greek 'onkos', meaning a lump).

Oncogenes were first detected in the genetic material of certain viruses that can cause infected normal cells to become malignant. (It should be said at this point that viruses are thought to be implicated in only a small number of human cancers). Then came the startling discovery that virtually the same base sequences could be found in the DNA of many cells *not* infected with the viruses. It has gradually become clear that these oncogene sequences are very similar to part of our *normal* DNA. By 1983 oncogenes had been detected on fifteen of the twenty-three pairs of human chromosomes. It is now believed that those few viruses with the ability to cause cancers had 'borrowed' oncogenes from mammalian cells at some period in their past evolution and had incorporated the base sequences into their own genetic material.

The discovery of oncogenes has led to a revolution in the biological description of cancers. The dominant explanation now incorporates the view that sequences very like the oncogenes were probably active in the embryo, or at later stages of development when the cell divides, but are inactive at other times. Cancers develop when these genes are inappropriately 'switched back on' as a consequence of them becoming slightly altered. In one instance, this alteration has been shown to involve the mutation of a *single* base in the DNA sequence within the gene.

□ The idea of switching genes on and off was introduced in Chapters 2 and 4 of this book. Can you describe the biochemical events that take place when a gene is active?

■ The base sequences in the DNA comprising the gene are used as a template against which a precisely corresponding sequence of RNA bases are built. This strand of RNA then leaves the nucleus, becomes attached to ribosomes somewhere in the cell and is used as a template against which a precisely corresponding sequence of amino acids are linked together to form a particular protein. Thus the 'code' in the DNA base sequences is 'translated' into a particular protein.

It is a huge jump from this rather mechanical analysis to an understanding of *how* the proteins produced when oncogenes are activated can transform the cell into a malignant state. A few clues are emerging, however. Perhaps the most interesting is the discovery in 1983 that two of the known oncogenes produce proteins that are virtually identical in amino acid sequence to certain so-called growth factors. Normal cells will not divide unless particular proteins (growth factors) bind to their membranes: for example, the dividing cell layers of the skin are rich in a substance known as epidermal growth factor, which is essential for continued cell division. Malignant cells have long been known to divide without apparent need for growth factors: it now seems likely that they can make their own. This may be part of the answer to an intriguing question of how so few oncogenes are able to regulate all the fundamental cellular changes associated with malignancy. The implication is that some of the proteins coded for by oncogenes could, in turn, switch on numerous other genes that are normally active only in embryonic cells. Therefore, a few oncogene proteins could 'orchestrate' the biochemistry of the entire cell.

In addition to discovering the effects of oncogenes on cells, it is essential to work out how they become activated and whether they can be switched off again. Very little is known about this to date, but there is compelling evidence that the activation of more than one oncogene is necessary before a cell becomes truly malignant. Thus, one or more oncogenes may control cell division, but others are probably involved in migration, loss of differentiation, adhesiveness and contact inhibition that characterises malignant cells. This fits epidemiological data which show that certain cancers are more likely if several risk factors coincide: for example, lung cancers are more common among cigarette smokers, but commoner still among people who smoke *and* drink significant amounts of alcohol. Similarly, while a virus is implicated in the cause of cancer of the cervix (the neck of the womb), cigarette smoking and use of oral contraceptives also appear to be risk factors. Certain environmental 'insults' (chemicals, radiation, some viruses) may turn out to activate different oncogenes and current research predicts that the cell becomes malignant only when several such genes have been switched on. The gradual accumulation of activated oncogenes may explain the rising incidence of cancers with age.

Precisely how a chemical or a dose of radiation could activate a gene is uncertain, but in some cases it may be by breaking the DNA chain close to the oncogene. Repair mechanisms exist in cells to rejoin breaks in the chain, but sometimes they go wrong and pieces of DNA end up being spliced into the wrong location, or even into the wrong chromosome, similar to the translocations that can give rise to congenital conditions such as Down's syndrome. A few types of cancer have already been found in which oncogenes appear in an unusual place in the DNA sequence, particularly close to the break point where part of one chromosome has snapped off and been exchanged for part of another one. Small changes in the sequence of DNA bases close to the break might be enough to activate the oncogene. Alternatively, the oncogene may be suppressed by nearby sequences of DNA in its original location and removed from their influence when the DNA is rearranged.

Oncogene theories focus on the activation of *pre-existing* sequences of DNA bases. Although oncogenes are currently the 'flavour of the month', an earlier notion still remains a possibility — namely, that factors in the environment (chemicals, radiation) could cause such damage to human DNA that significant alterations in some base sequences result and these 'new' portions of DNA mediate the malignant transformation. Either way, both theories imply failure of the repair mechanisms existing in all cells to monitor the correct DNA sequence and accurately mend breaks in the chain. The deterioration of these repair mechanisms with age may also contribute to the fact that cancers become more common as people get older.*

---

* In this discussion we have focused on the proximate causes of cancers — the genetic and biochemical changes that form the final step in malignant transformation. Distal causes are discussed in *Experiencing and Explaining Disease*. The Open University (1985) *Experiencing and Explaining Disease*, Open University Press (U205, *Health and Disease*, Book VI).

Summary
As you have seen in this chapter, the proximate causes of the majority of the conditions resulting in death, illness and disability in industrialised countries are unknown. As was indicated earlier, we have omitted one group of conditions which is responsible for much ill-health.

☐ Which conditions are these?

■ Mental disorders.

Our knowledge of the aetiology of mental disorders is even poorer than that of the conditions discussed in this chapter.*

It is possible that some of the conditions covered in this chapter will turn out to have a single proximate cause but, at present, it seems more likely that their aetiology is of a multifactorial nature. You have seen that the factors involved are many and varied: genetic, chromosomal, infectious, auto-immune, environmental (including diet, chemicals, radiation, climate, smoking) and psychological. In other words, they include all the factors that have been discussed as proximate causes of disease in the preceding four chapters. It should be remembered, however, that before single proximate causes were discovered for many of the conditions covered in these chapters, doctors and scientists had sometimes developed multifactorial theories to explain their occurrence. There may therefore be several surprises in store as the cause of coronary heart disease and stroke, cancers and arthritis are unravelled in the years ahead.

Objectives for Chapter 16
When you have studied this chapter, you should be able to:

16.1 Describe the pathology of a number of conditions including spina bifida, coronary heart disease, stroke, dementia and rheumatoid arthritis.

16.2 Describe the interrelationship of the following processes: atherosclerosis, thrombosis, ischaemia, infarction, cardiac arrhythmias, and heart failure.

16.3 Discuss some of the current ideas on the proximate causes of cancers.

Questions for Chapter 16

1 (*Objective 16.1*) Children with spina bifida often suffer from repeated infections of the urinary system. Why do you think these infections occur more often in these children than in 'normal' children?

2 (*Objective 16.2*) Describe the likely sequence of pathological events between the development of atherosclerosis in the coronary arteries and heart failure.

3 (*Objective 16.3*) Imagine that a single cell in the centre of a mass of 'normal' tissue undergoes malignant transformation. Many weeks later, a secondary cancer develops at some distance from the original transformation. Describe the characteristic behaviour of malignant cells during the journey from the primary cancer to the secondary one.

* Current ideas on the cause of some mental disorders are discussed in *Experiencing and Explaining Disease* (U205, Book VI).

17
Biomedicine: help or hindrance?

During this chapter you will be referred to two articles in the Course Reader, one by Ivan Illich (Part 3, Section 3.7) and one by Thomas McKeown (Part 3, Section 3.1). You will also be asked to read the article by Paul Beeson (Part 3, Section 3.3).

Throughout the last five chapters, we have included examples of diseases that have been, or still are, caused by doctors or other health-care professionals. Such diseases are known as *iatrogenic* (from the Greek words 'iatros', meaning physician, and 'genesis', meaning origins).* *Iatrogenesis* may take three different forms: clinical, social and cultural iatrogenesis. Clinical iatrogenesis comprises all conditions for which remedies, physicians or hospitals are the cause.

□ What examples of clinical iatrogenesis have been cited in this book?

■ You may have recalled the harmful effects of X-rays and some drugs (such as thalidomide) on the developing fetus; the effect of antibiotics in disturbing the commensal bacteria in the large intestine and the promotion of fungal infections; and the adverse reactions some people experience to drugs to which they are hypersensitive.

There are countless other examples of iatrogenesis. No branch of medicine is free of causing some harm to some patients. Why should this be, when doctors set out to help and improve the health of their patients? The answer seems to be that it is an inevitable consequence of the use of powerful methods of intervention, whether that be the use of drugs, lasers, ionising radiation, surgery or even psychotherapy. In Chapter 13, for example, you saw how harmful to health lasers may be and, yet, used under control in a medical setting, they offer a highly effective means of treating serious eye conditions that would otherwise lead to blindness. Although medicine makes every effort to minimise the serious long-term effects of such interventions, it seems unable to prevent them altogether. Indeed, as regards the more immediate adverse side-effects of many drugs and procedures, it appears that their prevention is

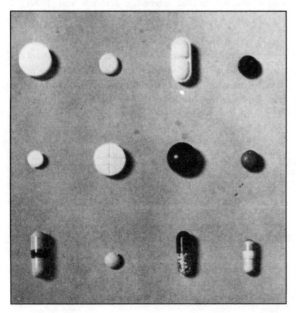

Pills, tablets and capsules — a representative sample from the thousands available.

* *Iatrogenesis* is the subject of an article in the Course Reader by Ivan Illich, 'The Epidemics of Modern Medicine'.

impossible. This is because of the nature of the doctor–patient relationship, and the *placebo effect*.

☐ Do you know what the placebo effect is?
■ Placebo comes from Latin and means 'I shall be pleasing'. The placebo effect refers to the benefit that a patient receives from a medical intervention that does not arise from the actual chemical properties of the drug, say, or physiological effects of the operation, or any other procedure they undergo.

In other words, the benefit appears to arise simply from the experience of the consultation or the stay in hospital. The reason for the placebo effect is the subject of much debate and research. It appears to be related to the patient's attitude to the care they are receiving, their confidence and trust in the doctor and their belief in the power of the procedure to which they agree to subject themselves. There are countless examples of the placebo effect, ranging from the use of drugs and electroconvulsive therapy, to surgical operations. Everyone is familiar with placebos — most of us have felt the benefit of cough medicines, many of which do not, in fact, contain any drug with a known pharmacological effect of helping alleviate a cough.* Thus, as long as the medical intervention from which the benefits of the placebo effect arise is not harmful in other ways, this state of affairs is to be welcomed.

So what has this to do with a discussion of iatrogenesis? Just as a pharmacologically inactive preparation, such as cough medicine, may produce beneficial, placebo effects, so it may also result in harmful, adverse side-effects. These have been called *nocebo effects*, though as yet this term has not gained widespread use. We should not be any more surprised by nocebo effects than by placebo effects, as they both arise from the same set of attitudes and beliefs about the power of medicine. The importance of the nocebo effect to a discussion of iatrogenesis is that, if the nocebo effect is ever present in medical interventions, then however careful doctors may be in reducing the harm arising from the 'active' ingredients of drugs, investigations or operations, the harm arising from the nocebo effect will still remain. This can be removed only by ceasing to practice medicine! In other words, as soon as a patient meets a doctor, the interaction will produce both beneficial placebo effects and adverse nocebo effects. Thus a world free of iatrogenesis is incompatible with the existence and practice of medicine.

Awareness of iatrogenesis is not a recent phenomenon,

as can be seen in this quote from Plinius Secundus, written in the second century AD:

> To protect us against doctors there is no law against ignorance, no example of capital punishment. Doctors learn at our risk, they experiment and kill with sovereign impunity, in fact the doctor is the only one who may kill. (from *Naturalis Historia*, 29.19, quoted in Illich, 1976, p.29)

There have been examples of iatrogenesis throughout the history of medicine, in particular during the two eras of therapeutic nihilism — the first during the early part of the nineteenth century, the second during the past twenty years.* The latter has, so far, culminated in the work of Ivan Illich, who claims that the degree of harm caused by doctors actually exceeds the benefits derived from their work. While this view represents an extreme position, there is considerable evidence supporting the more moderate position that doctors have made either no, or only a slight, contribution to the improvements in health that have occurred in industrialised countries.†

☐ To what can most of the improvement in the health of the British population since the eighteenth century be attributed?
■ Improvement in social conditions: nutrition, sanitation, clean water and other public health measures.

The view that medicine has played only an insignificant role in improving health has been based on changes in *mortality* rates. In other words, both Illich and McKeown have tended to base their criticisms of medicine on its failure to reduce mortality and extend life expectancy. This is only a limited view of the role of medicine. Other people have argued that medicine should be judged in terms of its contribution to the reduction of *morbidity* and *disability*.

In this book you have seen the extraordinary improvements that have taken place in our understanding of the structure and function of cells and intracellular chemistry, the physiology of the human body and the complex biological defence mechanisms, such as the immune system. In addition, you have read about the extent of present-day knowledge of the causes, pathology, pathogenesis and natural history of some diseases. Can it really be true that all this understanding of the biology of both normal and abnormal functioning has brought about only insignificant improvements in the health of humans? As you might expect, the exponents of biology and biomedicine would deny such a pessimistic suggestion. They argue that the beneficial results of their endeavours

* Some examples are given in *Medical Knowledge: Doubt and Certainty* (U205, Book II).
† This view is associated with McKeown whose work is referred to in *The Health of Nations* (U205, Book III).

*Pharmacology is the science of the action of drugs on living organisms.

Table 17.1 The meaning of some medical terms used in the article in the Course Reader by Paul Beeson

ectomies	medical slang which refers to surgical operations in which some tissue or organ is removed, e.g. tonsillectomy (removal of tonsils), cholecystectomy (removal of gall-bladder)
cytotoxic drugs	drugs that damage or kill cells, used in the treatment of some cancers
pharmacotherapy/chemotherapy	therapy which involves the use of drugs
internist	an American term for a physician, a doctor who specialises in adult internal medicine, such as the care of diabetes or pneumonia
orthopaedic	medical speciality concerned with muscular and skeletal abnormalities and disorders
haematology	medical speciality concerned with disorders of the blood
biopsy	the removal of a small piece of tissue from the body for detailed laboratory investigation
symptomatic treatment	a treatment that relieves symptoms but does not cure the underlying disease

are clear for all to see. This view can be seen in the article by Paul Beeson in the Course Reader (Part 3, Section 3.3). Beeson, until his recent retirement, was one of the leading physicians to practise in the UK and USA. In the article, the names of several diseases, with which you will probably not be familiar, are mentioned. Knowledge of these terms is not necessary to understand the main arguments Beeson uses to defend biomedicine. However, there are some other medical terms, the meanings of which are shown in Table 17.1, that you may find helpful. When you have read the article, answer the following questions.

☐ According to Beeson, what are the main changes that have taken place in the medical treatment of disease since 1927?

■ A reduction in the number and percentage of treatments that were harmful, useless, of questionable value or merely alleviated symptoms; an increase in those treatments that are highly effective, whether by curing or preventing disease.

☐ Do you think that Beeson's claim that many modern treatments are 'never likely to be classed with the bleeding and purging that characterised 18th and early 19th century medical therapy' is justified?

■ They may be but, as Beeson himself points out in discussing the limitations of his survey, 'Judgment of the worth of a given treatment ... is more likely to be faulty with regard to the present than when aided by retrospection after a half a century.'

The tendency to judge contemporary medicine as being a notable improvement over the past, and free of the mistakes of past generations, is a belief that has been held at all times. This view, the so-called Whig view of medicine, holds that, although mistakes may have been made in the past, these have now been left behind as humanity heads onwards and upwards into a bright new future. In support of this, Beeson

cites many examples in which there is evidence that current treatment is more effective than previous medical practice. In addition, simply ceasing to carry out harmful procedures has brought about some improvement in people's health, although as you have already seen, the dangers of iatrogenesis still exist.

What has been the contribution of biomedicine to improvements in health: major contributor, insignificant factor or harmful intruder? These three views are not as incompatible as they may at first appear. As has already been indicated, judgement of biomedicine should probably be based on its effect in alleviating suffering and disability rather than in reducing mortality rates. Viewed in this way, biomedicine can be seen as having been both a major contributor (in reducing morbidity) and an insignificant factor (in failing to reduce mortality). The accusation of causing harm has been a justifiable claim throughout the history of medicine and is supported by current levels of iatrogenic disease. Iatrogenesis, however, is an inherent feature of any medical system, whether it be Western biomedicine or any other.

☐ Beeson attributes the improvements in medical practice to an important change in the attitude and approach of doctors in their decision as to what treatments to use. What has that change been?

■ The acceptance and adoption of scientific evaluation of treatments, rather than basing decisions on 'unwarranted prejudices' based on personal experience.

Whether Beeson's confidence that all medical treatments will be based on scientific evaluation in the future is justified is a matter for debate. While such a prospect is to be welcomed, it is important to recognise that, in the 1980s, the majority of current biomedical procedures and treatments have still to be evaluated by the rigorous criteria that are currently deemed to be necessary to satisfy scientific objectivity.

Objectives for Chapter 17

When you have studied this chapter, you should be able to:

17.1 Explain what is meant by iatrogenesis, placebo and nocebo, and why iatrogenesis is a feature of any medical system.

17.2 Describe the contribution that biomedicine has made to improvements in health in industrialised countries.

Question for Chapter 17

1 (*Objectives 17.1 and 17.2*) 'Today's physicians must steel themselves to affirm in coroner's courts that treatment with effective modern drugs very often implies a small calculated risk of more or less serious side-effects — and that the risk is consciously and tacitly accepted because it is enormously outweighed by the much greater chance that the drug will restore health or actually save life.' (Miller, 1973, p.6)

    (a) What is the author's attitude to iatrogenesis?

    (b) Do you think this attitude is justifiable?

References and further reading

References

ASHBY, ERIC (1978) *Reconciling Man with the Environment*, Oxford University Press.

DUBOS, R. (1953) 'The germ theory revisited', Lecture delivered at Cornell University Medical College, New York, 18 March. Cited in S. Wolf (1981) *Social Environment and Health*, University of Washington Press.

ILLICH, IVAN (1976) *Limits to Medicine*, Marion Boyars.

MILLER, HENRY (1973) *Medicine and Society*, Oxford University Press.

THE CHEMIST AND DRUGGIST (1898) *Diseases and Remedies*, London.

WRIGHT, THOMAS (1971) *The Passions of the Minde in Generall*, University of Illinois Press. First published 1604.

Further reading

ASHBY, ERIC (1978) *Reconciling Man with the Environment*, O.U.P.

This short book discusses the political and scientific aspects of assessing environmental risks to health and the slow evolution of public conscience, from indifference to concern for the environment, over the past hundred years.

BRITISH MEDICAL ASSOCIATION (1983) *Report of the British Medical Association's Board of Science and Education: The Medical Effects of Nuclear War*, John Wiley.

An authoritative review of all the medical and scientific evidence of the consequences of nuclear war on human health, other species and the environment in general.

CLARK, BRIAN (1984) *The Genetic Code and Protein Biosynthesis*, Edward Arnold.

A detailed discussion of the structure of DNA and the mechanisms by which the code contained within this structure is translated into instructions for building proteins.

GOULD, STEPHEN J. (1980) *Ever Since Darwin: Reflections on Natural History*, Penguin.

Sparkling popular essays by a leading evolutionary theorist.

HARDY, RICHARD (1983) *Homeostasis*, Edward Arnold.

A straightforward guide to the regulation of the internal environment through nervous and hormonal pathways, and their interactions in complex control systems.

MEDAWAR, PETER (1984) *Pluto's Republic*, O.U.P.

The author is one of the most respected contemporary biologists. This book is a collection of essays ranging widely over scientific controversies.

MEDAWAR, P. B. and J. S. (1978) *The Life Science*, Paladin.

A beautifully written, though sometimes dense, general introduction to modern biology.

MESSENGER, J. B. (1979) *Nerves, Brains and Behaviour*, Edward Arnold.

A look at the kinds of information that animals collect about their external world and the relationship between behaviour and the organisation of the brain and sense organs.

MILLER, HENRY (1973) *Medicine and Society*, O.U.P.

This book is concerned with the impact of the revolution in biomedical science on society and medical practice. It deals with the present state of medicine in industrialised countries, the historical background of present developments and some possibilities for the future.

PUTNAM, R. J. and WRATTEN, S. D. (1984) *Principles of Ecology*, Croom Helm.

An account of the ecological perspective, though not specifically concerned with health and disease.

ROITT, IVAN (1984) 5th edition, *Essential Immunology*, Blackwell Scientific.

A lively, readable account of the fundamental principles of immunology including a detailed discussion of the genetics, biochemistry and cell biology of immune processes in health and disease.

ROSE, STEVEN (1974) *The Conscious Brain*, Penguin.

An introduction to the brain and nervous system's structure, function, development and evolution.

ROSE, STEVEN (1979) 2nd edition, *The Chemistry of Life*, Penguin.

A clear and readable introduction to biochemistry and cell biology.

ROWLAND, A. J. and COOPER, P. (1983) *Environment and Health*, Edward Arnold.

This book describes those elements of the environment which are known to influence human health. The authors discuss the ways in which the relevant environmental influences are measured and assessed.

SACKS, OLIVER (1984) *A Leg to Stand On*, Summit Books, New York.

A striking autobiographical account by an English neurologist of his own neurological problems.

THOMAS, LEWIS (1975) *The Lives of a Cell*, Bantam Books.

An account in praise of the potential and real value of basic biological research in the conquest of disease.

TOTMAN, RICHARD (1984) 2nd edition, *Social Causes of Illness*, Souvenir Press.

Reviews and critically examines the research into the psychological determinants of health and disease, covering such aspects as the impact of stress, the effect of personality and the role of emotions in the causation of disease.

YACOB, FRANCOIS (1974) *The Logic of Living Systems*, Allen Lane.

A fascinating history of biological thought by a leading French molecular biologist.

Further References for Chapter 15

DATTORE, P. J., SHONTZ, F. C. and COYNE, L. (1980) 'Premorbid personality differentiation of cancer and non-cancer groups: a test of the hypothesis of cancer proneness', *Journal of Consulting and Clinical Psychology*, vol. 48, pp. 388–94.

DEROGATIS, L. R., ABELOFF, M. D. and MELISARATOS, N. (1979) 'Psychological coping mechanisms and survival time in metastatic breast cancer', *Journal of the American Medical Association*, vol. 242, pp. 1504–8.

EVANS, N. J. R., BALDWIN, J. A. and GATH, D. (1974) 'The incidence of cancer among patients with effective disorders', *British Journal of Psychiatry*, vol. 124, pp. 518–25.

GREER, S. and MORRIS, T. (1975) 'Psychological attributes of women who develop breast cancer: a controlled study', *Journal of Psychosomatic Research*, vol. 19, pp. 147–53.

GREER, S., MORRIS, T. and PETTINGALE, K. W. (1979) 'Psychological response to breast cancer: effect on outcome', *Lancet*, ii, pp. 785–7.

HOLMES, T. H. and MASUDA, M. 'Life change and illness susceptibility', in DOHRENWEND, B. S. and B. P. (eds) (1974), *Stressful Life Events: Their Nature and Effects*, Chapter 3, pp. 45–72, John Wiley.

HOLMES, T. H. and RAHE, R. H. (1967) 'The social readjustment rating scale', *Journal of Psychosomatic Research*, vol. 11, pp. 213–18.

KERR, T. A., SHAPIRA, K. and ROTH, M. (1969) 'The relationship between premature death and affective disorders', *British Journal of Psychiatry*, vol. 115, pp. 1277–82.

KISSEN, D. M. (1963) 'Personality characteristics in males conducive to lung cancer', *British Journal of Medical Psychology*, vol. 36, pp. 27–36.

LAUNDENSLAGER, M. L. *et al.* (1983) 'Coping and immunosuppression: inescapable but not escapable shock suppresses lymphocyte proliferation', *Science*, vol. 221, pp. 568–70.

MORRIS, T. *et al.* (1981) 'Patterns of expression of anger and their psychological correlates in women with breast cancer', *Journal of Psychosomatic Research*, vol. 25, pp. 111–17.

NIEMI, T. and JÄÄSKELÄINEN, J. (1978) 'Cancer morbidity in depressive persons', *Journal of Psychosomatic Research*, vol. 22, pp. 117–20.

PURCELL K. *et al.* (1969) 'The effect of asthma in kinds of experimental separation', *Psychosomatic Medicine*, vol. 31, no. 2, pp. 144–64.

THOMAS, C. B., DUSZYNSKI, K. R. and SHAFFER, J. W. (1979) 'Family attitudes reported in youth as potential predictors of cancer', *Psychosomatic Medicine*, vol. 41, pp. 287–302.

WEISMAN, A. D. and WORDEN, J. W. (1977) *Coping with Vulnerability in Cancer Patients*, privately printed, Boston, Massachusetts.

WHITLOCK, F. A. and SISKIND, M. (1979) 'Depression and cancer: a follow-up study, *Psychological Medicine*, vol. 9, pp. 747–52.

WOLFF, C. T., *et al.* (1964) 'Relationships between psychological defenses and mean urinary 17-hydroxycorticosteroid excretion rates. Part 1: A predictive study of parents of children with leukaemia', *Psychosomatic Medicine*, vol. 26, pp. 576–91.

Answers to self-assessment questions

Chapter 1

1 The properties common to *all* living organisms are (b), (c) and (e).

2 Ecology studies the cat in its environment: how it relates to other cats and to other species (the mice and birds on which it preys) in the same terrain as itself (holistic).

Cell biology and biochemistry explore the structure of the cells which make up the cat, their molecular composition. They try to explain 'how the cat works' in terms of these molecules (reductionist).

Developmental biology explores the ways in which the kitten develops from the embryo inside its mother's womb and how the cat grows from a kitten.

Evolutionary biology explores the paths of descent of the cat from earlier mammals and the ways in which natural selection has shaped and adapted the cat to its present lifestyle. (These last two are historical sciences.)

Chapter 2

1 Proteins have a variety of roles in the body. Examples of major groups are the enzymes, which catalyse chemical reactions, and structural proteins, which provide the framework from which most of the body is constructed. Another protein mentioned in this chapter is haemoglobin, the oxygen-carrying chemical in the blood. Nucleic acids are the major carriers of information in cells, both in cell division and in the control of cellular processes. Carbohydrates are used mainly as chemicals in energy transfer reactions in human cells.

2 In the simplest terms, enzymes are proteins 'built' from a string of amino acids joined together in a particular sequence. The 'code' that determines the sequence for each protein is contained in a portion of DNA known as a gene. The code is 'written' in the form of four different bases in that portion of DNA. A group of three bases (a triplet) specifies a particular amino acid. The DNA code in the gene is first transcribed (copied) by constructing a molecule of messenger RNA which contains an exactly complementary sequence of bases to those in the DNA. This molecule of mRNA travels to structures called ribosomes in the cytoplasm of the cell and there the enzyme is constructed by linking amino acids together in the order indicated by the order of bases in the mRNA.

3 The sketch below identifies the various structures.

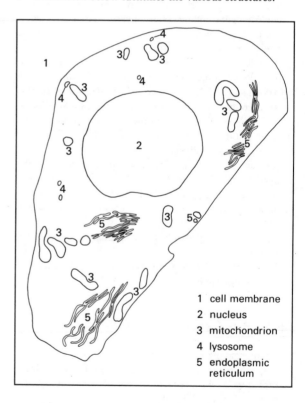

1 cell membrane
2 nucleus
3 mitochondrion
4 lysosome
5 endoplasmic reticulum

Chapter 3

1 (i) A nucleus is possessed by protozoa, but not by viruses and bacteria. (ii) Bacteria and protozoa have cytoplasm; viruses do not have a cell body. (iii) Only protozoa have chromosomes. (iv) Some bacteria possess a cellulose membrane, viruses have a protein coat and protozoa have a cell membrane composed of lipid and protein. (v) All three possess nucleic acid.

2 The kidney is an *organ* composed of a number of different tissues — muscle, epithelium and connective tissue. The kidney is part of the urinary system.

3 (i) Epithelium makes up the outermost layer of the skin (shown as (a) in Figure 3.6). Its function is to protect the body from external factors that might cause damage. Epithelium also lines the glands of the skin (shown as (e)). Some of these cells are specialised to secrete a substance which helps to maintain the surface of the skin.

(ii) Connective tissue makes up the bulk of the lower layer of skin, the dermis (see (b) in Figure 3.6). This is loose connective tissue. Its function is to support the blood and lymph vessels, nerves and hairs. Fatty (or adipose) connective tissue lies beneath the dermis (see (c) in Figure 3.6) and helps to insulate the body against excessive heat loss or gain.

(iii) Both striped and smooth muscle are present in the skin. Each hair in the skin has a minute striped muscle attached to it which can alter the position of the hair (see (d) in Figure 3.6). It is these muscles that can literally 'make your hair stand on end'. Smooth muscle is present in the veins and arteries beneath the dermis (though not in the capillaries).

Chapter 4
1 Each gene is made up of a limited number of bases arranged in a particular sequence, and there are only four possible bases at each base position along the gene. In fact, only a minority of possible base changes result in proteins being produced, either because 'nonsense' triplets are generated or because the chain of amino acids produced on the mutated gene does not correspond to any useful protein. Thus, although there are theoretically a very large number of possible base changes, only a very small number of observable *variations* in each gene actually exist in nature.

2 The possible combinations are as follows:

Parents: aa   AA

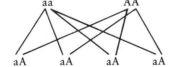

Possible children: aA   aA   aA   aA

That is, all children would be *aA* and, as the albino allele *a* is recessive, no child would be albino.

However, if the mother were *Aa*, the following children could result:

Parents: aa   Aa

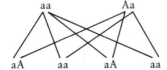

Possible children: aA   aa   aA   aa

That is, half the children would have alleles *aa* and so could be albino. The others would be *Aa* and would appear normal, but would be carriers of the albino allele.

3 (a) It resembles the cell of the 16-cell embryo in having the same general cell structure of nucleus, cytoplasm, and cell membrane. It contains an identical set of chromosomes and DNA to the embryonic cell. (b) It differs in that it has a characteristic shape; different amounts of its DNA are expressed in the manufacture of proteins; it is fixed in relationship to other cells rather than free-moving; and it has lost the ability to divide.

4 Cell division involves replication of DNA, and the copying of DNA bases is subject to error. Although a few errors are of little significance to an organism and may be repaired, the accumulation of errors can lead to impairment of cell function.

Chapter 5
1 The muscle cell and the bacterium both need oxygen and certain nutrients, particularly glucose, amino acids and salts such as those of sodium and potassium. They both need removal of the waste products of digestion and respiration. The bacterium is bathed in the external environment, and thus can obtain essential molecules and expel waste ones either by diffusion or active transport across its cell membrane. The muscle cell is totally surrounded by other cells and, even though there is a little tissue fluid between adjacent cells, this would rapidly become exhausted of nutrients and saturated with waste unless constantly replenished. Thus a transport network is necessary to bring nutrients from the digestive system and oxygen from the lungs, and remove waste. This network must be capable of pumping fluids around the body so that supply and waste removal is continuous. The biochemical reactions in the muscle cell also work most efficiently at a particular temperature (around 37 °C), so its temperature is kept fairly constant around this optimum; the temperature of the bacterium drifts with that of the environment and hence cannot maintain peak biochemical efficiency.

Chapter 6
1 (a) False. Although the statement is true for *some* hormones (particularly thyroxine and growth hormone), it is not true for all. Some hormones affect specific target tissues, producing a range of responses, many of which do not directly affect the utilisation of food for energy (that is, alter the metabolic rate).

(b) False. Although hormones produce sustained effects over longer time periods than the rapid bursts of activity in the nervous system, they show marked fluctuations with varying periodicity ranging from minutes to years.

(c) True.

(d) True.

(e) False. Although all endocrine glands receive signals from the nervous system, few are directly under its sole control, and some glands secrete hormones predominantly under stimulation from other hormones.

2   The additional doses of growth hormone would cause the rat to grow abnormally large, even though maintaining normal body proportions. The excess growth hormone in the bloodstream would inhibit the pituitary gland from producing growth hormone of its own (the 'overshoot' would be fed back to the gland, shutting down its production — this is an example of a negative feedback loop).

3   (a) Sympathetic part of the autonomic nervous system.

(b) Brain stem or midbrain.

(c) Several parts would be involved: the eyes and optic nerves, the visual area in the cerebral cortex, and memory.

(d) A reflex arc involving the peripheral nerves and spinal cord.

Chapter 7

1   (a) The artery carrying blood from the heart to the lungs contains deoxygenated blood, and the vein returning blood from the lungs contains oxygenated blood. In all other arteries and veins in the body, the state of oxygenation is the other way round.

(b) Arteries do not have one-way valves to help them transport blood against the force of gravity, unlike veins which do this in most parts of the body. Therefore, although each pulse of the heart pushes blood up the arteries of the arm, it tends to flow back down between contractions.

(c) The thickness of arterial walls is mainly so that they can withstand the high pressure of blood leaving the heart. Veins require only thin walls because they carry blood at low pressures, and for the same reason must have a relatively wide diameter to cut down resistance to blood flow.

2   (a) Dilating all veins would produce a sudden drop in blood pressure, since the (unchanged) volume of blood would be held in a larger volume of blood vessels. Lower blood pressure would result in a more sluggish return of blood to the heart and hence in lower heart rate and cardiac output.

(b) A sudden increase in blood *volume* increases blood *pressure* and hence produces a temporary increase in heart rate and cardiac output.

Chapter 8

1   Compare your answer with the completed version of Figure 8.5 opposite.

2   The baby and the elderly person are less able to generate heat than an active child or adult, partly because they cannot move around so easily, so you would expect their body temperature to fall. However, young babies have much less ability to regulate their temperature than adults of any age, so their temperature would fall faster. Drinking hot liquid would help to maintain the body temperature of both people in this experiment but, whereas the elderly person could excrete all the excess liquid, the baby could not. The baby's blood pressure would rise with the increase in blood volume, but the kidneys could not increase urine formation by very much. Fluid would collect in the tissues, and the 'drinking centre' in the hypothalamus would be inhibited so that the baby stops drinking. The elderly person, even though not thirsty, could over-ride the hypothalamus and drink the hot liquids as a conscious activity to keep warm. All in all, the baby is much more vulnerable than the elderly person.

Chapter 9

1   (a) Hydrogen peroxide is an important defence against bacteria contributed by cells involved in non-specific immunity. In its absence, chronic bacterial infections result (particularly in the skin), but these are rarely fatal since the antibodies and cytotoxic cells of the adaptive immune system are still produced, and phagocytic cells can engulf bacteria.

(b) Total absence of antibodies leaves the individual vulnerable to most bacterial infections, in particular because antibodies are needed to 'label' suitable targets for phagocytic cells, and also because they trigger the cascade of enzymes (known as complement) which can break down bacterial cell walls. The immune response to viruses depends mainly on the direct killing of infected cells by T lymphocytes and chemicals that prevent viruses from replicating, so immunity to virus infections remains reasonably good.

(c) Without T cells, the entire adaptive immune response breaks down. Not only are cytotoxic T cells missing, but B cells cannot produce antibodies because there are no T cells to give them a signal to do so. Thus the only aspect of immunity that remains intact is the contribution of macrophages and other phagocytic white blood cells, but this is not adequate because in the absence of antibodies, potential targets have not been 'labelled'. Children born with this defect die very quickly from infection unless a successful thymus graft can be performed.

2   (a), (iii); (b), (ii); (c), (ii); (d), (i); (e), (i); (f), (ii).

Sentence (f) illustrates how a product of the adaptive immune response (antibodies) can enhance the phagocytic ability of cells involved in non-specific immunity.

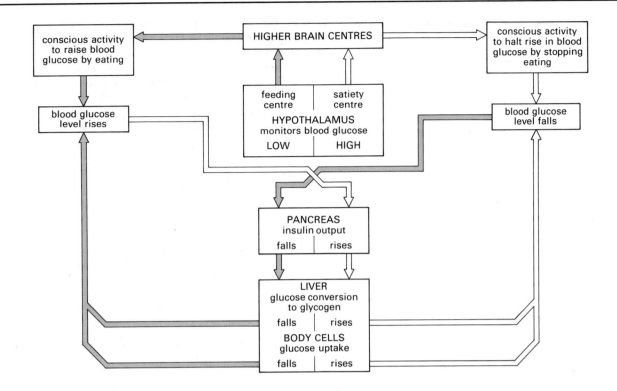

3 Vaccines contain relatively harmless preparations of the organism against which protection is needed. The unique antigens on the organism are recognised by the relatively small number of lymphocytes that are committed (in advance of meeting the antigen) to respond to that antigen alone. They respond by undergoing repeated cell divisions, rapidly increasing their numbers and eliminating the antigen. After the response dies down, some of this clone (i.e. the population of identical cells) survives, and forms the basis of a more effective and rapid response if the antigen is encountered again. This occurs after the second immunisation, which leaves the clone larger than it was before, and so on.

If exposure to live organisms becomes likely years after immunisation, then a single dose is given shortly before the probable exposure. This not only reactivates the appropriate clone, but also ensures that a high level of the relevant antibodies and cytotoxic lymphocytes are already circulating in the body *before* the organism is encountered. An infection may then never have a chance to develop.

Chapter 10

1 (a) False. Androgens are the hormones responsible for primary sexual development. All fetuses are exposed to high oestrogen levels because they develop inside the mother, so oestrogens would not be useful for producing sex differences.

(b) False. It is the sex hormones that are directly responsible for the physical changes of puberty. These hormones are, however, produced in response to changes in pituitary hormones (gonadotrophins).

(c) False. At birth, the ovaries contain the entire stock of ova (although in an immature state), whereas the testes produce new sperm throughout adult life.

(d) True. Although the sex chromosome of the fertilising sperm determines the sex of the offspring (since all ova are X whereas sperm can be X or Y), *which* sperm fertilises an ovum is affected by the chemical conditions in the female's reproductive tract.

2 (a), (v); (b), (iii); (c), (i); (d), (iv); (e), (ii).

Chapter 11

1 Ecology is concerned with the interrelationship between organisms and their environment. The ecological view regards disease as the result of the breakdown in the state of harmony that usually exists between an organism and its social and physical environment.

2 (1) The effect of natural selection diminishes as an individual grows older, so that it is not within the power of

natural selection to prolong the lengths of human lives. (2) Human evolution is ultimately bound up with the evolution of other species, some of which may cause disease and death in humans. The evolution of a disease-free, immortal state in one species is incompatible with an ecological view of disease, which requires all species to be taken into account.

3   (a) Viruses and bacteria; (b) smoking, poor social conditions, a family history of bronchitis; (c) damage to the bronchi; (d) sudden onset, repeated attacks.

Chapter 12

1   Each protein made by cells has a different function. If an abnormal protein is manufactured by many cells in the body and takes place in key chemical reactions, or forms part of many structures, then its abnormality may have widespread effects.

2   You should have noted that the condition occurs only in men, and that it is passed on through normal mothers, but not through normal or affected fathers. These facts suggest that the gene for colour blindness is a sex-linked-recessive gene, on the X-chromosome. Men are affected but their sons cannot be. They do, however, pass on the defective X-chromosome to their daughters, who are then carriers. Men who are colour-blind *always* inherit it from their mothers. If you are confused about this, study the following two possibilities of inheritance:

Either

Parents:	xY		XX	
	affected man		non-carrier woman	
Sperm and ova:	x	Y	X	X
Possible children:	xX	xX	XY	XY
	female carriers		unaffected sons	

Or

Parents:	XY		xX	
	normal man		carrier woman	
Sperm and ova:	X	Y	x	X
Possible children:	Xx	XX	xY	XY
	carrier female	non-carrier female	affected male	unaffected male

Chapter 13

1   (a) Obesity, from over-consumption.

(b) Deficiency diseases in certain sub-groups of the population, such as osteomalacia from insufficient vitamin D and anaemia from inadequate iron intake.

(c) Disorders resulting from the diet, in particular, the fibre, fat and sugar contents.

(d) Disorders caused by chemical contaminants.

2   The bone will heal by the formation of new cells from the surface layers of the broken ends. The final result (after several weeks) will be indistinguishable from the old bone, assuming it is suitably treated. The muscles cannot heal by cell division, and will therefore form scar tissue. This will lead to tightening and contraction of muscles, which in the long term may result in deformity of the arm and a restriction of its range of movement.

3   (a) Some constituent of cigarette smoke, thought to be a tar, may damage DNA in the nuclei of cells in the lungs and this may give rise to cancer.

(b) Irritants in the smoke may cause excessive mucus production in the lungs. This may lead to damage to the bronchi, resulting in chronic bronchitis. Damage to the cells lining the alveoli results in emphysema.

(c) Carbon monoxide may combine with haemoglobin in the red blood cells, leading to a reduced capacity to carry oxygen around the body. This could be serious in someone already suffering from an impaired circulation, for instance, as a result of atherosclerosis. (The development of the latter is also promoted by cigarette smoking.)

4   (a) Damage might be caused by a lack of iodine in the child's mother's diet during pregnancy. This most frequently occurs in people living in mountainous regions and where the rainfall contains little iodine. As a result, there are only low levels of iodine in the soil and thus in the food they consume.

(b) Noise. Boilermaking is one of the 'noisiest' occupations, and unless the ears are protected, damage to hearing will almost certainly occur.

Chapter 14

1   Whether the *Streptococcus* gets into the host's throat and multiplies will depend on the immune competence of the individual, and the virulence of the strain of the bacteria. Those two variables mean that *Streptococcus pyogenes* may or may not cause disease when present in the throat.

2   Colds and influenza are spread in the air, and this is not easy to prevent, whereas typhoid and cholera are spread by faeces contaminating drinking water, prevented in this country by good sanitation and a clean water supply. Two other reasons why typhoid and cholera are also easier to control than colds and influenza are: (i) they are caused by bacteria and can be treated by antibiotic drugs (though, in fact, these do not help much in cholera). Colds and 'flu are caused by viruses against which there are no drugs; (ii) there are reasonably efficient vaccines against typhoid and cholera (though these are not used since public health measures are adequate). Widespread immunisation against colds and influenza are not effective because of the multiplicity of causative agents (colds) or their changing

antigenic properties (influenza).

3 (i) Their means of transmission — foodborne, waterborne, and direct contact with faeces — can be adequately broken only by good sanitation and a clean water supply, neither of which are available in most areas of the Third World. (ii) As yet, no effective vaccines against worms have been developed. This is partly because of the 'masking' or absense of antigens on the surface of most worms. Also, the main response to worms is inflammatory rather than immune. (iii) Although there are several effective drugs available for the treatment of worm infestations, people do not acquire any lasting immunity to worms. Successful eradication by drugs will almost certainly be rapidly followed by re-infection. It therefore makes little sense to treat worm infestations unless they are causing significant distress or morbidity.

4 The antibiotics, which are aimed at killing off bacteria in the throat, lungs or wherever the infection may be, also disturb the commensal bacteria in the large intestine. The antibiotics will kill the most susceptible species and leave the more resistant ones to flourish and cause diarrhoea. In practice, the normal balance between the different species is usually restored in a matter of days.

Chapter 15

1 There are two features that suggest the involvement of psychological factors. The first is the nature of the 'effective treatment' for the asthmatic attacks — being comforted by his mother. It would seem unlikely that an allergy to house dust or some other factor would respond to such treatment. Second, the sudden cessation of attacks when he no longer had his mother at hand to comfort him. You may also have considered whether the attacks in childhood were to gain the attention and comfort of his mother. However, with no further information about this relationship, it is possible only to speculate.

2 No, not necessarily. It might be that during times of family crisis, children spend more time with relatives, childminders or schoolfriends and are hence in contact with many different people, some of whom might have streptococcal throat infections.

Chapter 16

1 Spina bifida often affects the nerves running from the spinal cord to the bladder and other parts of the urinary tract. Damage to these nerves may result in loss of nervous control of the bladder so that urine may continually dribble. In this situation, infection can ascend into the bladder, and beyond to the kidneys.

2 *Atherosclerosis* will deprive the heart muscle of sufficient oxygen. If the supply of oxygen is cut off completely, as a result of a *thrombus* forming on an atherosclerotic plaque, then a *myocardial infarction* will occur. The death of an area of cardiac muscle may reduce the efficiency of the heart, by causing a *cardiac arrhythmia* and this will result in the *heart failing* to cope with maintaining the circulation.

3 The primary cancer enlarges as a result of the unrestrained division of malignant cells, which undergo mitosis more often than nearby 'normal' cells. Loss of contact inhibition and an increased ability to move enables some malignant cells to squeeze between adjacent normal cells, and slowly migrate through the surrounding tissue. Enzymes may aid this process by breaking down the 'cement' between cells. When a malignant cell penetrates a capillary of the blood or lymphatic systems, it is swept away downstream and may eventually lodge in a fine capillary network some distance away. There it may begin to multiply by repeated cell divisions and form a secondary cancer (metastasis).

Chapter 17

1 (a) Iatrogenesis is an acceptable risk for doctors to take with their patients lives if they believe the treatment might 'restore health or actually save life'. (b) It is justifiable if there is scientific evidence to show that the risks of a treatment are 'enormously outweighed' by the benefits. Although such evidence exists for some medical treatments, there are many that have never been scientifically evaluated.

Acknowledgements

Grateful acknowledgement is made to the following sources from material used in this book:

Figures

Figure 3.6 after Plate 104 (Integument Organization) from *The Anatomy Coloring Book* by W. Kapit and L. M. Elson, copyright Wynn Kapit and Lawrence M. Elson, reprinted by permission of Harper and Row, Publishers, Inc.; *Figure 9.1 (a)* reproduced with permission from N. J. Bigley *et al. Immunologic Fundamentals*, 2nd edn., copyright © 1981 by Year Book Medical Publishers Inc., Chicago; *Figure 9.1 (b)* Dr C. G. Cochrane in J. Bellanti, *Immunology*, W. B. Saunders, 1971; *Figure 9.3 (a) and (b)* courtesy of Drs S. Gordon and G. C. Macpherson, Sir William Dunn School of Pathology, University of Oxford; *Figures 9.4, 13.5, 13.6 and 14.6 (a) and (b)* courtesy of Royal Infirmary Sheffield; *Figure 9.5 (a) and (b)* from I. Roitt, *Essential Immunology*, Blackwell, 1977; *Figure 9.6* from N. K. Jerne, The Immune system, in *Scientific American*, vol. 229, © 1973 by Scientific American, Inc., all rights reserved; *Figure 12.5* MENCAP National Centre; *Figure 13.1* Professor G. C. Arneil; *Figure 13.2 and 14.11* Wellcome Medical Foundation; *Figure 13.3* S. Parvin; *Figure 14.2* Wellcome Museum of Medical Science; *Figure 14.4* C. James Webb; *Figure 14.7* Science Photo Library; *Figure 14.8* Dr S. G. Brown, International Leprosy Association; *Figure 15.1* from K. Purcell *et al.*, The effect of asthma in kids experimental separation, *Psychosomatic Medicine*, vol. 31, no. 2, 1969, Elsevier Science Publishing Co. Inc.; *Figure 16.2* ASBAH; *Figure 16.3* Department of Pathology, Edinburgh University; *Figure 16.8* Maclean Hunter Ltd., Medical Division; *Figure 16.9* A. C. Stansfield, St. Bartholomews; *Figure 16.10* G. Canti, St. Bartholomews.

Illustrations

p.vi from J. B. de C. M. Saunders and C. O'Malley, *Anotomical Drawings of Andreas Vesalius*, Harper and Row, 1982; *p.5* Bodleian Library; *p.35* The Director, India Office Library and Records (British Library); *p.49* Wellcome Institute Library, London; *p.54* reproduced with permission of Harper and Row from Olmsted, *Claude Bernard Physiologist*; *p.64* from B. Inglis, *A History of Medicine*, Wiedenfeld and Nicolson, 1965.

Plates

Plates I and III Open University Histological Laboratory.

Index

Entries and page numbers in **bold type** refer to key words which are printed in *italics* in the text.